HOMO DIGITALIS

Thierry Geerts

How digitalisation is making us more human

Lannoo

www.lannoo.com
Register on our website to regularly receive a newsletter with
information about new books and interesting exclusive offers.

Design cover: Herman Houbrechts
Layout: Studio Lannoo (Aurélie Matthys)
Illustrations: Ann-Sophie De Steur
Translation: Textcase

© 2022, Lannoo Publishers nv, Tielt, and Thierry Geerts
D/2022/45/230 – ISBN 978 94 014 8562 3 – NUR 740

Contents

The homo sapiens is no more. Long live the homo digitalis!

1

Digitalis in the 2020s

Why digitalisation will lead to more welfare

What are you afraid of? Great heights? Spiders? Thunder and lightning? People are fearful of a great many things – and that's perfectly normal. Similar to these fears, many people are also afraid of innovation. You might even lie awake at night thinking of the impact of technology on society, your job, or your relationships with other people.

It's an age-old fear. The 1920s are remembered as being one big party, the era of the Charleston, world's fairs, and beautiful cars, but it was also a time when people feared electricity, industrialisation, and wars. We're now in the 2020s, and, though we no longer fear past innovations such as the locomotive, we're now afraid of new phenomena like robots and artificial intelligence. We're afraid of losing our

← A century ago, we celebrated.

jobs to cyborgs, of the State becoming like Big Brother, and of losing our language and culture.

Fears like that can lead to populism, nationalism, and fascism – something that Hermann Göring knew all too well and exploited to help get the Nazis into power. According to the Nuremberg trials, he put it on record that it is actually relatively easy for a politician to get the support of the people by playing into their fears.

The fear of technology is fed by insecurity. We don't know what the future might bring in this rapidly changing world. As Yoda from Star Wars taught us, fear is often the worst path to follow, but it's an interesting compass. After all, the fact that

Fear can lead to populism.

we're insecure and maybe even scared shows that we're not indifferent to our fellow human beings and the society we live in. The fears we have reveal the things we find important: meaning, privacy, nature, democracy, culture, etc.

Technology in itself is actually neutral. It provides both opportunities and risks. It's all about what we do with it, how we inform ourselves, how we rise above those fears and take matters into our own hands. That's what makes the difference and is the reason why this book's subtitle – 'How digitalisation is making us more human' – is an important statement. We're in control of the impact technology will have.

The past few centuries have showed us time and again that technology has helped us make great strides once we overcome our fear of it. We now live longer, work less, and are healthier than ever before,

The past few centuries have showed us time and again that technology has helped us make great strides once we overcome our fear of it.

thanks to the innovations that came into our lives over the past few decades. Yet, the aversion to new technologies remains persistent – a fear fed by being unavoidably confronted with negative elements of it on a daily basis, like when you are in a bar with your friends and someone takes out their smartphone instead of talking to you, or when rents rise in big cities because real estate moguls are buying homes en masse so they can rent them out as Airbnbs.

Ambient computing

These days, over 1.5 billion smartphones are sold every year. Europeans spend about 2.5 hours a day on their smart-

> The smartphone as an interface will also gradually disappear from our lives.

phones; half of them even say that their smartphones have become absolutely indispensable to their lives. Nevertheless, like many other technologies, the smartphone as an interface will also gradually disappear from our lives.

We won't forever be holding a device every time we want to search for information or look at something. Technology will increasingly be present all over the place but in a more practical and respectful way. The rollout of so-called ambient computing might turn out to be the most important revolution of the coming years: technology will be everywhere but will disappear behind the wallpaper. Smartphones will no longer take up so much time and space in our lives.

Today it's already possible to wake up in the morning and talk to smart devices, like Google Home or Alexa, that serve as a central point of contact for all devices in your home. Who would have imagined that ten years ago? They help you get the lights to provide you with just enough brightness in the morning, or you can ask your fridge to recommend recipes based on what you've got in there. In the United States, sixty million people already own one of these smart speakers. But that's only the first step of the revolution. Gradually, that device will also disappear as a central point of contact, and all electronic devices, from refrigerators and cars to doorbells and thermostats, will become 'smart' and you will be able to talk to them.

These things you see around your house will also come to the cities. At present, we can already see the first steps being taken into the frictionless world that ambient computing can provide us. For instance, in a lot of European countries people can now enter a public parking garage without taking

Ambient computing will ensure that technology will be everywhere.

a ticket. The parking garage knows who you are through licence plate recognition and the parking gate opens up; when you leave, the payment is settled automatically through your banking app. How convenient is this?

Smart glasses can scan the world around us and understand what we're seeing. They will allow you to walk into a Starbucks in China, see a tasty-looking cupcake, and order it without having to know a word of Chinese. Eventually glasses like these, much like our current phones, will also disappear and be replaced with an invisible lens of sorts.

That's not as crazy as it might sound: in the research centres of the world, hundreds of engineers are working on developing the technologies that will make such applications possible. Let's not forget that a smartphone would have been considered a strange idea just twenty years ago. Terms that often come up in such predictions are spatial computing, internet of things, 5G, edge computing, etc.

> Let's not forget that a smartphone would have been considered a strange idea just twenty years ago.

When reading the description above, you might have been thinking of Google Glass or the 'smart' refrigerator you saw in the electronics store – appliances that are mainly just expensive and clunky right now. We're still in the phase where technological gadgets are 'cool' but fall short because they're just not user-friendly enough. However, this will soon change, because ambient computing is getting better by the day.

Thanks to exoskeletons, we will be able to easily lift things that weigh over 200 pounds.

When smartphones disappear from our daily lives, peace and balance can also return. Even though technology will be present everywhere, it will be in the background and function in more of a supportive role, which will be beneficial to our (mental) health. I believe this will lead to harmony in the 'battle' between human beings and machines and that the latter will make the former better. We'll reach a sort of *augmented humanity*: humans will be able to rise above themselves and become more human with the help of technology. Sometimes this will be physical, such as exoskeletons that allow workers or nurses to easily lift things that weigh a couple hundred pounds. Sometimes this will be mental, with the internet functioning like an 'extra brain' or algorithms that can make our lives easier. Technology will take over less meaningful tasks so we can have more time for the things that really matter. However, this will largely depend on our own drive, passion, curiosity, and entrepreneurship. The opportunities are already there, but we'll have to learn to control the risks.

Reinvent the world

Innovations such as smartphones, artificial intelligence, and ambient computing are more than just gadgets. Just like steam engines, electricity, and the computer, they are leading to a new Industrial Revolution. The impact of this fourth Industrial Revolution should not be underestimated. It will force us to reinvent everything with the help of these new technologies.

We might say we're living in Denmark, Italy, or Germany, but this isn't really the case anymore. Together with four billion other people connected with each other online, we're living in a new 'country' which I christened 'Digitalis' in my previous book. That country is continuously evolving and goes far beyond the boundaries of the physical world. We can now watch series on American Netflix, post videos on Chinese TikTok, listen to music on Swedish Spotify, and buy products from all over the world with a single click.

We're now nearly all inhabitants of that one country I call Digitalis. But there is more. People create technology which, in turn, influences people. The changes on a technological, human, and social level have been so significant that people have evolved into what you could now call an entirely new species: the homo digitalis. The homo sapiens is no more. Long live the homo digitalis!

We might still look the same, but the way we live has completely changed. We now enrich our lives thanks to digital applications and explore the possibilities of Digitalis, with an instant connection to four billion people of the same species. This impacts our health, the way we learn, the relationships we enter into with other people, how we organise our lives, and so on. For the homo digitalis, a lot of boxes appear to have disappeared. There are new rules; we live differently, and everyone can become their own medium or brand. By now, there have been many books about these general changes, but not as many talk about what these changes are doing to us, something I am trying to map out in this book. At times I will have to be very brief when doing so, which is great, but it can also be a curse when writing a book: every chapter really ought to be a book in and of itself!

> The homo sapiens is no more. Long live the homo digitalis!

This fourth Industrial Revolution is a digital revolution.

The homo digitalis is a multimillionaire

Of course, people aren't standing still in Digitalis. Technology will always cause upheaval, sometimes driven by unforeseen circumstances. When the Covid crisis hit in 2020, the world discovered that many things could also be done digitally. Grandparents started using videoconferencing and online chat, and children joined their classes digitally. Digitalis was able to welcome a lot of new citizens. All of a sudden, working from home became the norm and people realised that even a visit to their doctor could be done via Zoom, Meet, or Teams and was even safer. Contactless

Homo sapiens has evolved into homo digitalis.

payments and e-commerce turned out to offer more secure solutions given the circumstances.

Digitalisation also accelerated in the workplace. The pandemic succeeded in ways that few managers have ever managed to do: digitalising organisations at record speeds.

Digitalisation can have far-reaching consequences for homo digitalis and our world. The best illustration of this is what happened ten years ago with the encyclopedia.

For centuries, the Encyclopedia Britannica was an institution. The 2010 edition counted over 32,000 pages and cost over $1,000. The thing that was so special about this edition is that it was the last to be printed after 250 years. Thanks to Microsoft Encarta and later Wikipedia, the paper encyclopedia, weighing 130 lbs., wasn't as valuable anymore.

The decision to stop publishing the book had a tremendous impact because the encyclopedia no longer needed to be physically compiled, printed, or distributed. Thousands of trees were saved, as was the fuel previously used for the trucks distributing the books. But it also negatively impacted the economy and the gross national product.

However, many more people were now able to access the knowledge since it had become so much cheaper to do so. Microsoft Encarta costs less than $100 and Wikipedia is completely free! Four billion people now have access to very useful, always up-to-date, high-quality encyclopedias for free.

Digitalisation makes certain things a lot more affordable and therefore more easily accessible. Try to calculate how much the information and services provided on your smartphone would have cost around 1990. Think about everything your device can do: it's a camera and a compass, a complete encyclopedia, and it can access more videos than any video store could ever have stocked. A rough estimate is some $30 million, most of which is the computing power. Our current phones possess more computing power than all of the Apollo space programme! So basically, the homo digitalis is a multimillionaire, though it doesn't feel like it. In fact, we more often denounce technology than see it as a valuable

All of the books in the world can fit on your smartphone.

asset. Maybe it's best to not always focus on the economic aspects but instead broaden the discussion.

There are an increasing number of voices both in countries and companies saying we need to focus more on well-being than purely on economic growth. The two don't always go together and we sometimes lose sight of human well-being. Nevertheless, well-being is a better compass to use when mapping out digital policies and developing the right future mindset.

There are different ways to map out people's well-being and the resulting happiness. For instance, in Bhutan, they focus on the gross national happiness measured against specific parameters such as health, time usage, culture, and social cohesion. There are also other essential parameters that are of importance, such as safety, given that this is one of the most critical conditions when it comes to being happy. Another factor is sustainability, which, though often forgotten, will ensure that we can remain happy for decades to come.

With all of these parameters in mind, I am taking a look at the changes happening in Digitalis, the country that connects over four billion people via the internet. I am building on the things I talked about in my first book, mainly in an effort to consider the impact the changes are having on the citizens of Digitalis. Whereas in my first book, I mainly looked at the broad trends, I will now examine how these trends are changing our lives.

We homo digitalis are curious beings. In this book, you'll discover, among other things, that we may possess less than the homo sapiens, but also share more. By spending less time on repetitive tasks, the homo digitalis is given more time for profundity, creativity, care, and personal development. We live in Digitalis, a world where the focus is increasingly being placed on well-being and high programme. But is this leading to greater or less happiness?

Based on my findings, I am daring to claim that we can be happier than ever before and will live in a world that's better than it ever has been. Whether it makes you happy or not is something only you can decide, but I'm absolutely convinced that our level of well-being will rise.

Maybe you're more likely to dwell on the problems caused by technology. In recent years, it has turned out that digitalisation has sometimes amplified certain dangers and even created entirely new ones. The Cambridge Analytica scandal did, for instance, create a political shockwave and recent elections in the United Kingdom and the United States have shown the consequences of fake news.

In the following chapters, I will first focus on the homo digitalis itself, after which I will zoom out and look at the changes happening in the world the homo digitalis inhabits. In both parts, I'll argue in favour of technology, not dismissing the opportunities it will provide us with and discussing how we can manage the risks.

At the end of the book, I'll focus on what we can do to get on track. And by 'we', I'm talking as much about you, your company, or your employer as I am about our governments.

Yes, Europe can definitely play a leading role in Digitalis: it's not too late.

We homo digitalis can be happier than ever before.

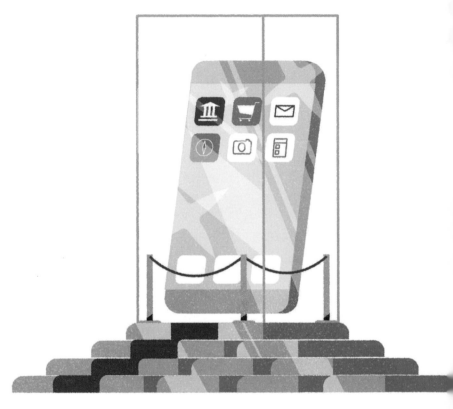

Now sacred, but perhaps soon it will be a thing of the past:
the smartphone.

Now that the new decade has begun, it's high time to deal with the issues at hand and look ahead into the future, to conduct constructive and thorough conversations, to learn how to manage the risks, and to seize all of the opportunities offered by digitalisation. Homo sapiens are ingenious creatures. What type of future will they face?

2

Who is the homo digitalis?

Will we all become cyborgs?

The humans of the future will look suspiciously similar to us. The homo digitalis will most likely not be some kind of superbeing or 'cyborg' like the ones seen in science fiction movies. After all, today's humans aren't really all that different from the humans of eight centuries ago. To them, however, we would undoubtedly come across as 'cyborgs', since we wear lenses and keep our hearts ticking with pacemakers, two innovations in healthcare that have drastically improved our lives. Even more is in store for the homo digitalis.

← The digital revolution is raging and has a significant impact on us as humans.

2.1   Healthcare for the homo digitalis

Can we improve our healthcare system and keep things affordable?

It was better than ever before, or so we thought. Two world wars, a cold war, and a series of economic crises hadn't been able to break us down. We thought we were untouchable until one minuscule virus showed us in a matter of weeks that our entire society relies on a healthy working class and relies on ensuring our healthcare system is not overloaded. One spark was enough to set the entire thing on fire.

The Covid crisis took on a technological dimension soon enough as people began discussing how apps and other digital tools could help combat the spread of the virus. And, indeed, the relationship between the healthcare and technology sectors is one that will only strengthen over the coming years – not just in times of crisis, but also, and maybe especially, in our daily lives.

A cabinet member once asked me which sector should absolutely embrace digitalisation. My response: undoubtedly the healthcare sector. Only by digitalising it will we be able to grow old with dignity and keep everyone healthy in an affordable way. A healthier population brings economic advantages because we can keep the costs of ageing within limits. But what will the concrete impact of digitalisation be?

How will doctor visits change?

Three major evolutions will take place in the short term. The first one might just be the most boring, but it will save the homo digitalis a lot of trouble: technology will simplify administration. Without us fully realising it, a lot has changed in that regard behind the scenes in recent years.

It used to be that you had to call your doctor in order to make an appointment. After the appointment, you would be asked to pay and then to wait for the insurance to settle things in order to get a refund a few months later. If you needed to see another doctor, you almost always had to retell your entire medical history to get the new doctor up to speed.

Once fiction, now reality: algorithms and robots that help doctors.

Luckily, you can now make appointments online with most doctors and your health insurance will eventually be able to settle the bill automatically. Electronic patient records now allow every provider in your network to view your medical history, making it easier and more reliable the next time you have to see a specialist.

A second change is the rise of remote medical visits. After all, it's absurd that we think it is normal for sick people to go to the doctor's office and share their germs with everyone else in the waiting room. To avoid these situations, homo digitalis will come to trust remote doctor's visits more. Since Covid, we have seen a surge in consultation via video chat and everything points towards this becoming the new norm.

Of course, a medical doctor can't always make a complete diagnosis from a distance, but it's often possible to make an initial diagnosis. A doctor can't conduct every test through telemedicine, but a patient can measure their own temperature, for instance.

However, when we do go to the doctor, it would make more sense to always complete a questionnaire or submit our biometric data in advance. Think about the Migraine Buddy app: those who suffer from frequent migraines can log their migraines and thus possibly help correlate them to their diet, fatigue, stress, or other causes. The patient can then discuss this with the doctor, who also has access to the log.

This will allow doctors to talk to patients about how they really feel and the doctor can now focus even more on the human aspect. Soon they will be assisted by artificial intelligence to do an even better job, which will lead to a third evolution in the field of medicine. This sounds a lot worse than it actually is. When hearing the term AI, you might spontaneously think of all-knowing digital beings, but, first and foremost, artificial intelligence is a technology that has no 'mind' of its own nor the ability to engage in critical thought. However, it is very capable of recognising patterns.

> Doctors will be assisted by artificial intelligence to do an even better job.

This can be very useful in, for example, breast cancer research, a disease that has already seen tremendous progress in its treatment thanks to AI. It's not always easy for radiologists to see whether or not a suspicious spot is cancerous or not based on a mammogram. The probability of a false positive (women are told they have breast cancer while this is not the case) or a false negative result (the cancer is overlooked) is no less than *one in three*! Those misdiagnoses lead to a lot of medical and psychological suffering.

By studying over 100,000 anonymised mammograms, an algorithm was developed by Google that is now capable of getting a more accurate result than a team of doctors. The number of misdiagnoses went from one in three to fewer than 2%! Radiologists have rightfully called such algorithms 'the most revolutionary invention since the scanner'. They will never replace the doctor but are an extra tool that will ensure a much more accurate diagnosis.

Similar algorithms are also being used in poorer countries to prevent the spread of diseases that have largely disappeared from the Western world, such as diabetic retinopathy, which results from poorly controlled diabetes and can cause blindness.

Radiologists call such algorithms 'the most revolutionary invention since the scanner'.

In countries such as India with a chronic shortage of specialised ophthalmologists, almost half of patients suffer a visual impairment before they are even diagnosed. With the help of artificial intelligence, healthcare providers can determine whether or not a patient is suffering from diabetic retinopathy based on a scan of the retina. This test can even be conducted by support staff so that a lot more people can be tested.

The administrative simplifications, the rise of telemedicine, and the changing relationship between doctor and computer are just the initial wave of changes to healthcare, but of course, it won't stop there. The homo digitalis is wearing a Fitbit, smartwatch, etc., mainly as a pedometer or portable messaging device so far, but these devices will have a huge impact on our health in the future.

An app a day keeps the doctor away

Those smart devices are a first step in the shift from 'curing' to 'preventing'. In a perfect world, you'd rather pay your doctor to ensure you don't get sick. Today, you pay them for treatment because your ailment wasn't prevented. You can already notice this in the language we use: we go to the hospital to get 'better' or 'treated'.

> Those smart devices are a first step in the shift from 'curing' to 'preventing'.

Smartwatches will evolve into medical devices that will be perfectly capable of keeping track of our biometric values. At first, this will just be fairly general figures such as our

Smartwatches will evolve into medical devices.

heart rate or blood pressure, but others will soon be added. In Japan, the company Toto is experimenting with a smart automated toilet monitoring and analysing urine and stool.

By combining personal data with international epidemiological data, one could prevent infections from certain diseases instead of contracting them. But things become even more interesting for people with conditions such as diabetes. These patients have long had to stick themselves in the finger to get a blood sample to measure their sugars, but luckily there are now painless alternatives available. For instance, there is a sensor that can be adhered to the upper arm that injects a tiny needle underneath the skin to measure blood glucose levels. The biggest disadvantage is that this smart device needs to be switched out every two weeks. Innovative companies such as Indigo Diabetes are looking into solutions for this and developing a chip the size of a fingernail to be placed invisibly under the skin of the abdomen. This small device then communicates with a receiver the size of a usb stick, which then transfers the data to a smartphone. The results are then used to notify the patient if their sugar levels are off.

It's an important innovation that will improve the lives of millions of people. The homo digitalis who suffers from diabetes will soon be able to take it one step further and wear an insulin pump that, based on the data, can automatically make additional injections if necessary.

The fact that healthcare in Digitalis will be more data based than ever is a given. Today, services like 23andMe are

trying to map out hereditary diseases based on a saliva
sample. There is still a ways to go in regards to accuracy, but
it is setting the course for the future. You can for instance get
a better understanding of your genetic risk related to obesity
or your predisposition to certain addictions.

This is a very sensitive matter, especially because nothing
is certain. Just because you're not predisposed to obesity
doesn't mean you can eat a Big Mac every day. Ethical issues
also play a role: to what extent do you want to know which
hereditary diseases you might have? I think your life could
be hell if you knew the details of what might lie in your
future, which is why it's crucial that such information is first
assessed by a doctor. The doctor can then decide whether
and how the results should be shared with the patient.

That same doctor is also playing a role in another important shift from 'getting healthy' to 'well-being'. While we
used to focus mainly on the
physical aspect, the homo digitalis will adopt a more holistic
approach. This approach to
well-being used to be reserved
only for top managers or athletes who had a personal coach
to help them work on their mind-body balance.

There is another important
shift from 'getting healthy'
to 'well-being'.

When we are better able to map our biological values,
medications could be more personalised than is currently
the case. Nowadays, the pharmaceutical industry works as a
mass industry by standardising all its products. Anyone who

is sick goes to the pharmacy and buys the same products as everyone else.

I myself have struggled with high blood pressure for years and I go annually to the doctor to have my blood pressure checked. Based on that one measurement, I'm then given a prescription for the rest of the year. That's insane, of course, because there is always a chance that certain factors could have influenced the results on the day of my visit.

Inhabitants of Digitalis will be called by their doctor at certain times. The doctor will tell them they see in the readings that the person's blood pressure has been rising for a while and will immediately have personalised medicine delivered to their home.

The impact on healthcare will be huge: maybe I've waited a few years too many to start taking those pills. Thanks to future innovations, I will be able to immediately adjust my dosages close to perfection and it will give me a healthier life. In order to develop these types of personalised medicines, very powerful computers will be needed to perform the calculations required and digitalise production processes. The good news is that we're almost at the point where such things are becoming technically feasible.

Many lives can be saved in this way because the current industrial approach of the pharmaceutical and medical sectors has a lot of negative consequences. For instance, there is a big difference between male and female bodies, but treatment options are all too often still only tested on male subjects. There are many female-specific conditions for which no medicine has ever been developed and the same dosages get incorrectly prescribed to both men and women.

By combining machine-learning algorithms with the knowledge we possess about the human genome, we can begin to trade the industrial approach to medicine for a more personalised system. This will also allow the pharmaceutical sector to reposition itself. Currently, the industry often struggles with a bad reputation of greed, even though they are doing important work by keeping people healthy. If they don't soon realise that turning a profit should not be their primary concern, there is a good chance that the sector will not survive for long. Caring only about profits might have been an option during the Industrial Age of pharmaceuticals, but in the digital era, it will become increasingly difficult to attract talent by doing so. If this were the case, there is a chance that the pharmaceutical companies will have to deal with disruptive newcomers, much like what happened to the taxi sector with Uber.

Finally, Digitalis will also provide the opportunity for greater international cooperation and exchange of information. This turned out to be an absolute necessity during Covid as we quickly remembered that diseases and epidemics don't care about national borders. By mapping out online information on a global scale, more targeted actions can be taken and the speed at which vaccines are developed can be accelerated. When it comes to storing such sensitive information online, a lot of issues surrounding privacy and security will arise, of course, but that is something which I will address later.

The fact that we'll live healthier in the future is one thing, but how we live our daily lives will also drastically change. Homo digitalis will use digital gadgets such as smartwatches and algorithms to live a better and more efficient life. But will it really work?

2.2 Efficiency in the daily life of the homo digitalis

Do we really need all of these digital gadgets?

When we talk about the impact of technology, we often talk about the things that can be digitalised, but it would be better to turn that sentence around and ask which things can't or shouldn't be digitalised. You'll soon end up with human qualities such as creativity, empathy, collaboration, and taking context into account, as well as matters related to physical needs such as food, drink, or clothing. How far we should take things is something that's up for debate, but every chapter of this book discusses elements of life that could be handled even better to increase our quality of life. Practical things can be done faster, better, or more easily, allowing more time to deepen our understanding, creativity, self-development, and general well-being.

It's a common thread throughout this book and in *Digitalis*, which was published a few years back. Of course, my predictions were, at times, rather optimistic in terms of timing. For instance, like many others, I thought that completely self-driving cars would be available much sooner. The truth lies somewhere in the middle: self-driving cars will be around both sooner and not so soon. The futuristic vehicles that will be able to take us from point A to B completely on their own are already being rolled out in a pilot project by Tesla and Waymo, but it will take years before we will all

We enjoy the benefits of connected homes, cities, and mobility.

have access to them. Nevertheless, cars that take over parts of the tasks of drivers are already finding their way into our garages. For instance, after writing *Digitalis,* I tested the amazing Tesla Autopilot. Many other cars, even those in lower price ranges, now have adaptive cruise control and lane assist.

You could say the same thing about teleportation, a dream many people in Hollywood have been playing with for decades. The chance of people being able to teleport from one location to another is low (but never say never), but maybe we won't have to. We're now able to transport ourselves as pixels, which is sufficient most of the time. Through video chat and hyper-realistic virtual reality in the near future, you can (virtually) be on the other end of the world in no time.

Step by step

It's a classic mistake to lose sight of our ultimate goal. For instance, in the previous chapter, I wrote about personalised medications, but it may take years for this to become common practice. Yet there are already an increasing number of tools, both in plain sight and behind the scenes, that are helping to keep you healthy in a personalised way.

Gadgets and applications that might seem superfluous at first glance might be able to significantly impact the way we live, transport, handle our finances, shop, or spend our free time in the future. It's only after looking back on things that we notice all of the small steps we've already taken. Take the digital success of the Belgian banks, for instance, where more than half of customers no longer even visit the physical bank anymore. Belgians are now using digital banks daily.

Over 2 million Belgians are filing their tax returns online.

We can open accounts or transfer money while sitting on the couch. It wasn't all that long ago that we had to go to the bank to do so or at least log on to our PCs.

The same can be said about our tax returns. Over 2 million Belgians are filing their tax returns online. They generally do this by logging into the Itsme app (a digital ID card), a great example of a tool that is making the life of homo digitalis more efficient. It used to be that if you wanted to log in, you had to do that via a cumbersome EID card reader or the 'tray' at the bank. Now you can use your smartphone to identify yourself securely and, once logged in, it often turns out the information has already been filled in automatically. How convenient is that?

On the other hand, if you only partially implement this digitalisation, you end up in crazy situations. A great example was how we managed payments at the start of the Covid pandemic – people still had to type in their PIN when paying at the store, which created a great risk of transmitting the virus between customers. The crisis resulted in the breakthrough of contactless mobile payment like Apple Pay, Google Pay, and local banking apps. This will be the norm and typing the PIN will be a thing of the past.

Contactless payment was seen as quite strange at first, but it quickly became a habit as soon as the initial feeling faded. We discovered contactless payments out of necessity, but there is next to zero chance that people will switch back again to cash and card payments after the crisis. It might seem like a small step, but by now, we've already forgotten how revolutionary it was.

Will it really help you?

When there's a new kind of technology or application on the market, there will always be a phase in which it feels like a toy meant only for people with lots of money or tech geeks. People said the same thing when the first (expensive) GPS devices began to become widespread in the nineties. Back in those days, people usually had a paper book filled with road maps in their glove box! After that, Waze came out, which offers real-time updates and an opportunity to even talk to it! It was initially also considered an unnecessary app by many but is now being used by more than 130 million people.

The fact that many technological gadgets still feel a bit clumsy in their use is because they're still in their infancy phase. People who want to use a so-called smart assistant often need to give literal commands – 'Talk to H&M' – for specific actions to be performed, instead of keying in www.hm.com. This is getting better every day as the under-lying algorithms keep on learning, leading to a more natural interaction for the homo digitalis.

As mentioned in this book's introduction, it will mainly be ambient computing that will cause major upheavals. Anyone who wanted to do something online in the past had to walk to a computer to do so. Then we got smartphones, but they're also not all that practical. The next step will be the internet being built into the devices we use, which will then be 'smarter' and therefore

The next step will be the internet being built into the devices we use.

more useful and personalised, no longer just dumb gadgets. That will literally be the beginning of a new era.

It will also be an era in which many things we as humans have to do today will finally disappear. All of the tedious and repetitive actions you perform in your life – buying coffee every other week, proofreading spreadsheets at the office, emailing back and forth to plan an appointment – can now be automated.

How companies will better serve the homo digitalis

There are various sectors in which customers have benefited from automation. As I already mentioned, the banking sector might be the best example. When the financial crisis hit in 2008, no one was proud of their bank. At best, it was considered a useful necessity, similar to how energy suppliers are seen today.

More than a decade later, some of those banks are the best example of how digitalisation can work for everyone. Several people who aren't generally that tech-savvy have come up to me over the years and told me how pleasantly surprised they've been with the things their banking apps can do, whether it's splitting a bill at a restaurant or buying bus tickets.

> Banks have realised who 'the person behind the customer' is.

What these banks know how to do well is to serve their customers better, which can have surprising results. Nobody ever expected we would end up buying bus and train tickets directly from a bank app, but nowadays more tickets are purchased through the apps of our banks than from the public transport operators themselves in Belgium.

Banks have realised who 'the person behind the customer' is. By gaining more insight into the homo digitalis, they have been able to improve our lives and make them more efficient. You can see the same happening in other sectors now as well – just look at the car industry, where the brands learned that there were target groups that didn't need to own big cars (which often consume a lot of fuel). For instance, Citroën developed the Ami, a small electric car that doesn't have to be fast at all but is perfect for driving around town. Or look at Fiat, which focuses on car sharing with its Dream Garage: people who buy an electric Fiat 500 gain access to a garage with Jeeps and Maseratis they can borrow!

Everyone can now choose to feel like a race car driver for a day or borrow a bigger car to go on holiday.

The companies that have homo digitalis as a customer – from utility companies to supermarkets – will have to try their best to become more than just a supplier in a useful way.

Why isn't there an application from a utility company that can manage my entire house? An app like that could also help energy suppliers with their transformation into service companies. Everyone needs to decide how to heat their home at some point in their lives. We require ordinary citizens to become heating specialists overnight, but the energy supplier could also choose to supply 'heating at the right temperature' for a fixed amount.

Digital gadgets, from virtual assistants to incomprehensible algorithms, are indeed more than just gadgets. They're capable of drastically improving our quality of life. In this chapter, I have exclusively focused on the life of the homo digitalis, but, later on, I will also tell you more about how they impact the world we live in. These innovations will change our approach to life, leaving us with a lot of questions. Let's start with education.

2.3 Learning in Digitalis

Is the internet making us dumber?

Imagine that every baby born already has access to all of the knowledge humanity has ever amassed. A child will already know from birth that water boils at 100 °C because their father learned this in primary school. Or that they know the difference between left and right because someone on the other side of the world decided this centuries ago. And every baby born after this child will also automatically have access to that knowledge, including everything the child has learned in her or his life. It's a pretty strange and scary thought, isn't it?

Imagine that every baby born already has access to all of the knowledge humanity has ever amassed.

However, that's exactly how today's software works. When one of Tesla's self-driving cars drives into a pole in Oklahoma, the information is processed and added into the next update available to every Tesla in the world. If the AlphaGo Zero gaming console algorithm discovers a particular move has no result, this move will not be repeated in new versions.

This insight shows how the speed with which software learns is increasing exponentially – and that our human intelligence will one day be surpassed by that of technology. There has been a debate among experts for years about when that will happen. But virtually everyone agrees that it will happen someday.

You can adopt two attitudes when faced with this realisation. If you're more of a pessimist, you could become defeatist and resign yourself to the fact. Those who are more

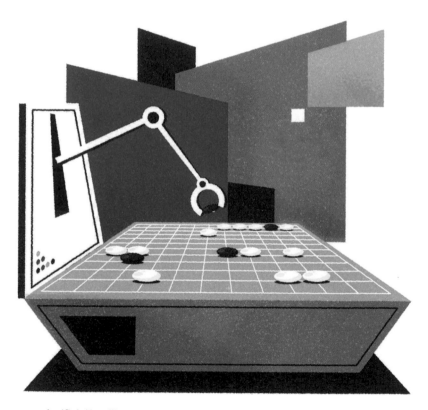

Artificial intelligence can handle even the most complex problems.

positive might just start thinking about how technology can help us deal with education and lifelong learning in the future since education will have to teach students how to deal with the risks and possibilities of these technological evolutions at an early age.

Parents already often complain about how their children live in a virtual world and don't spend enough time on schoolwork, but school is more abstract to children, almost like a virtual concept that has lost its connection to reality. Unlike the real world, they aren't allowed to use an app like Google Translate in school if they can't understand certain words. They're also still regularly expected to be able to recite factual knowledge, and there isn't any level of personalisation in play that would make the learning process more efficient. The things they're taught are sometimes far off base from the skills they will eventually need in real life. We learn about Napoleon and ancient Rome, but very little about how we should treat others, live healthily, understand our own emotions, or manage our finances and taxes.

That ought to change quickly in these two ways: we need to *learn with* and *about* technology.

> We will have to teach students how to deal with the risks and possibilities of these technological evolutions at an early age.

How we learn *with* technology

After the invention of writing and the printing press, the worldwide release of the internet has unleashed a third cultural revolution. However, 25 years after its development, the internet has not yet made it into every classroom. We instead use books, a medium that's been around for about four hundred years by now, and we just recently stopped using chalkboards, invented roughly a hundred years ago, to learn about today's world. Some people might want to mention the rise of smart boards, but these are often used just like digital chalkboards. Similar to how the first car was basically a carriage with a motor, it's merely a copy of an existing concept and not a revolutionary improvement.

> The worldwide release of the internet has unleashed a third cultural revolution.

It would already be an important first step to embrace the devices young people use every day in the classroom. This can be done by allowing them to look up information on their smartphones during history class or identify flora and fauna with the help of apps such as Audubon or ObsIdentify during school trips.

Language purists may say that allowing students to use applications like Google Translate in the classroom will make them develop less of a feel for the language, an unfounded fear in my opinion. The calculator hasn't made math class redundant; indeed, we've been able to raise the content of the lessons to a higher level because students no longer need to constantly do all the calculations in their head or on pen and paper.

The 2020 lockdown led to a quick technological revolution in classrooms. In some schools, teachers worked together for each subject: one of them taught two hundred students simultaneously through a video chat from a film studio while three other teachers answered students' questions, and a fifth teacher took care of those who needed extra help or to be challenged more.

At the same time, the lockdown exposed a lot of issues. Within a couple of months, we learned that a school has two functions: one educational and one social. The purely educational aspect can be replaced by technology surprisingly well, but the second aspect can't. The thing students missed the most during the lockdown was contact with their friends.

We also realised that being taught through a computer is very efficient but also exhausting. In an analogue classroom, students can daydream or stare at a bird on the playground, but being expected to stare at a screen for eight hours a day is tedious.

This way of teaching isn't always convenient for teachers either. We expect too much from teachers: they must have professional knowledge which they have to convey in an in-depth and compelling way, function as a pedagogue, sense which students need extra help, keep track of ever-increasing piles of paperwork, etc. The jacks-of-all-trades who can do all that are rare.

We need to unburden them and use technology for tasks that can be outsourced. My main thesis in this debate might also be the most controversial: much teaching might just as well be done through online videos. The majority of the theoretical classes taught are nearly the same every year and the way these classes are taught is often less exciting than we'd like them to be. A single person simply has less knowledge than the four billion people living in Digitalis combined. Teachers can now make one proper recording of their theory and then continue using it as a foundation for their following classes. The topic discussed in the recording can then be supplemented with other – and often better – classes that can be found online on platforms such as Skillshare or Udemy. By working in this way, students who think things are moving a bit too slow can skip certain things for themselves, while students who need more time can watch additional tutorial videos. Based on data collected from this process, a teacher could also find out which parts of the video are paused or rewound the most. Those parts could then be re-recorded until they'd eventually end up with the ideal lesson.

By combining the at-home theory lessons that students can follow at their own pace with practical lessons at school, especially in secondary and higher education, we can combat fatigue and raise the bar for students who can achieve this. Students want learning materials that will be relevant later in life. Think, for example, about lessons on mental health, personal finance, entrepreneurship, etc.

By doing this, teachers can also teach in a more targeted way so that the students' study hours are better spent. Students are now in school for seven to eight hours a day and then have to do at least an hour of homework or study for a test. That's a longer working day than most of their parents have!

In such a futuristic scenario, teachers will take on more of a coaching role. They will be the people trying to help students discover their passions and guiding them on their path, because it is that passion that drives people to keep on learning. Teachers make sure students like doing the work by teaching enthusiastically and/or finding interesting online examples. This way, young people are prepared for the real world.

Teachers know the strengths and weaknesses of their class like no other, allowing them to set the bar at the right level to provide the necessary challenges. As soon as students understand the usefulness of specific lessons and these lessons are translated in a way applicable to their own living environments, they will see their benefit.

Teachers will take on more of a coaching role.

Not only that, but when it clicks in students' heads, they will go looking for extra material online themselves. When I was young, children had to wait for something to be taught to them or they had to depend on their parents to teach them. With any luck, this might have sparked an interest in mathematics or languages, but, generally, children were less able to explore interests through their own initiative. Now it's the other way around and children are free to explore Digitalis. They can become experts without any help.

A great example of this can be found in my own house. My youngest daughter taught herself English in order to understand YouTube videos. Without ever having been taught the language, she managed to gradually understand what was being said thanks to those videos and now speaks better English than I do.

You should definitely take that 'coaching aspect' of teaching seriously: just like sports coaches, they can accurately analyse why certain students score poorly on tests based on

Schools are digitalising at breakneck speed, and teaching via videos is becoming more popular.

data such as the mistakes made. Perhaps some students always fail when a test is taken on a Monday and other factors come into play or others perform better when there is more time between the lesson and the test.

In some countries, more and more teaching is being done through video, which leads to education at a higher level being accessible to a larger group of people. In countries such as India, MOOCs (massive open online courses) have become very important. According to the UNESCO Institute for Statistics, the number of Indian primary schools that have access to electricity is increasing every year. In 2018, the last year for which data was available, more than half of the country's schools now had electricity, a first in the country's history. Those schools often had to make do with what they had, in terms of both material and the level of the teachers.

One of the scale-ups that plays into this is BYJU's in India, a MOOC platform worth over $10 billion that has been downloaded more than fifty million times. Lessons at BYJU's are still not cheap by Indian standards, but it's a first step towards democratic education.

We must also pay attention to the more vulnerable students who are at risk of falling by the wayside if education were to increasingly rely on computers or smartphones. The government plays a vital role in this because they need to realise that access to knowledge and the ability to use it will drive progress. Where oil-rich countries used to have an advantage over the rest of the world, the same will now apply to those that heavily invest in education. China has already realised this and is putting an unprecedented focus on educating its next generations.

Where oil-rich countries used to have an advantage over the rest of the world, the same will now apply to those that heavily invest in education.

The task of the government is therefore twofold: provide everyone with access to online classes and make teaching structures more efficient. The first could, for example, be done by arranging schools in such a way that students with difficult family situations can still watch the videos at school or by lending them the devices they need to watch the material at home.

How we learn *about* technology

I mentioned earlier that the talents and interests of students should be more central in education, which also impacts how we learn *about* technology. By letting them get familiar with some aspects of education in a playful way, we can make them enthusiastic about a sector that will have the most influence on their lives in the future.

It's essential to start this as early in life as possible so we can capitalise on children's inherent sense of wonder. The things a person considers 'new technology' are usually those innovations that appeared after the age of twelve. For example, people in their sixties see smartphones as an example of new technology, but these are just a part of teenagers' lives, while the latter group may consider vr glasses and smart fridges to be high-tech.

IT classes often only start when students are twelve, but girls have already hit puberty by that age, so they may have lost that childlike perception when they start the lessons. If IT lessons were to start at an earlier age, it's more likely that girls would also pick up an interest in those technologies and would want to continue learning about them later. And that's important, because now it's mainly boys that pursue careers in the IT sector, which means that the people who are developing our innovations don't represent the diversity of the society they are shaping.

The things a person considers 'new technology' are usually those innovations that appeared after the age of twelve.

Of course, initiatives such as CoderDojo – a coding education movement from Ireland – are already driving a positive change, but in my opinion, they mainly reach the parents of students who are already convinced. However, what they understand very well is that learning through play works. When schools talk about technology, they sometimes approach it very theoretically, as if they were discussing some sort of alien being. But as I have written before, it's often like the schools are an artificial concept to students. So just let the students experiment, because people learn by doing. When I was a child, my mother let me play with fire under her supervision so that I could learn the limits myself, instead of trying to teach me by scolding.

During lectures and interviews, I'm sometimes asked if we should teach everyone to program. That's easier said than done because technology evolves too fast to learn specific programming languages. I was still taught COBOL, Fortran, and Turbo Pascal, three languages that won't get you very far today. The media likes to push this perception that we should train everyone to become a programmer, which of course isn't the case. Students should first learn to understand how these devices work and how they can make the best use of them. Similar to how one learns about electricity in physics classes but not everyone is expected to become an electrician. So, if asked whether students should be taught a general understanding of programming, the answer is yes.

If asked whether students should be taught a general understanding of programming, the answer is yes.

Our education system is divided into three branches that are equally important: scientific courses with many subjects such as physics, mathematics, and chemistry; the humanities and social sciences; and the training courses in which people are taught to work more with their hands. All three will also be equally relevant in the future and will each feel the impact of technological developments. That's why it's equally important for plumbers and hairdressers-to-be that they receive lessons about this because there is a good chance that they will be preparing invoices with artificial intelligence or training colleagues via virtual reality in the near future.

A skill that's becoming more important is media literacy. These are the competencies needed to be able to deal critically with classic, new, and social media. These need to be taught from an early age because it's not always easy to distinguish fact from fiction in a world overloaded with information.

Education should provide context by discussing wrong or fake messaging at length. By looking for fake news related to subjects from the students' own environment, they'll quickly understand the system behind it. That way, young people can be educated to be critically thinking citizens – and that's much needed. It's only when young people discover where they can gather correct information that they will be able to participate in the democratic public debate and thus help shape society.

Today, there is no longer any distinction between one's life at school and the real world that starts as soon as you've got your diploma. No, from now on, it will be all the more important that we keep on learning our entire lives and continue to develop ourselves. Fortunately, technology will also be able to help us with that.

2.4 Digitalis is bringing out the human in us

Or are we becoming selfish, shallow, and aimless?

Close your eyes and imagine the future. Chances are that you envision a metropolis where cars are flying under a clear blue sky, a world with self-driving or self-flying vehicles, glass towers with lots of greenery. Our grandparents had the same vision of the future fifty years ago when they were asked what the world would look like in 2020.

When we talk about the future, we tend to focus on mobility, buildings, hardware, and things we can see. As people, we might be too obsessed with the physical, but these won't fix our problems.

We need to look beyond these changes because many of the things that will change our world are a lot less visible. Real innovation will be in the knowledge and interaction with the four billion (and soon: 8 billion) people in Digitalis and not just with them, but also with all of the connected devices that will make us smarter or help us. In order to learn how to use these, reap their benefits, and make wise use of the time gained, homo digitalis will need to focus on lifelong learning and personal development, because we won't be able to

Real innovation will be in the knowledge and interaction with the four billion (and soon: 8 billion) people in Digitalis.

learn everything through education. We can do this by learning from others because the internet forms a web of knowledge and connections.

Never graduate again

As I wrote before, the need for lifelong learning will become a more important factor in education. Students need to discover what they're good at in order to excel at it in the future, and they should at least acquire a certain level of basic understanding of other matters. I'm aware that some may consider this form of education too easy, but by moving away from that mindset, we can progress as a society. Rather quality over quantity, because the community thrives on people who excel, and by finding a job that matches our passion, we become a lot happier.

It used to be a lot harder to excel at something: if you only put information in books, you'd have to learn the entire scientific vocabulary of a language to become the best at something. Now you only need a basic vocabulary and, above all else, to be adept at using it effectively, because if you come across a word you don't know, Google Translate will help you. In my time, anyone who wanted to get an exemption from French to become a commercial engineer had to take a test in which they asked, among other things, what the word *éboueur* means. That didn't make any sense. It's not like you'd become a better commercial engineer just because you knew the French word for 'garbage man'.

Therefore, the focus must be even more on the use of language or mathematics to develop their talents more.

Knowing where your passion lies will become more important for everyone. We have shorter working weeks than before, but we will have to work for longer, just because we now live a lot longer. It will therefore be crucial to find a job that matches your interests and skills as closely as possible.

The chance that you can keep doing that same job for a long time with the same diploma is close to zero. Moreover, the likelihood that the job you'll end up doing ten years from now already exists is relatively small. A constantly changing world forces us to learn new things and develop new skills continuously. The stories of people in their fifties who lose their jobs because they're no longer deemed 'useful' and are then replaced by more digitally literate people are painful, especially because you know that these situations can be prevented. After all, it's a lot cheaper for companies to train their employees than to recruit and train new staff.

Some companies have realised this at some point and started organising refresher courses on digitalisation for all their employees. They didn't hire an external trainer for these courses to teach the employees in some classroom at a remote industrial park. No, they left the theoretical lessons behind and opted for training that focused on the jobs at hand. After all, digitalisation isn't a separate profession but a tool. The way this is approached varies from company to company, but my experience shows that it's best to combine the carrot (a promotion or bonus) with the stick (keeping the job).

Networking with your online network

Personal development is not always something you do on your own. You also learn thanks to the feedback of others – and this is one of the things Digitalis makes more accessible than ever.

In the past, the size of your circle of friends and acquaintances depended on the budget you had for stamps or phone calls. Over time, the contact with these people faded, leaving your circle quite limited. Now you can communicate with anyone at any time – I could only dream of having those opportunities in my youth.

Imagine being fascinated by hummingbirds back in the eighties. You would count yourself lucky if you could find just one book about these birds in your local library. It also was a lot more challenging to find people who shared your enthusiasm. For instance, I am interested in the dangers of populism, but my friends sometimes want to talk about other topics when we go to a bar.

Digitalis can bring all those people with a shared passion into contact with each other. One of the most popular websites in the world is Reddit, a collection of online forums about specific topics. There are over a million different forums, or subreddits, in total. For example, r/hummingbirds brings together 4,200 people from all over the world who are

Imagine being fascinated by hummingbirds back in the eighties. You would count yourself lucky if you could find just one book about these birds in your local library.

passionate about hummingbirds, and over 170,000 fans of bonsai trees have found each other on r/Bonsai.

These communities discuss their common passion and share articles. They give each other feedback and tips, stimulating their personal development in the subject. The knowledge they find there eventually goes way deeper than that one book from your local library.

When we talk about Digitalis, many only see the vast amount of content that is offered. They talk about an endless stream of apps, sites, and online platforms that provide distraction. People often ignore how in-depth the knowledge can be online: the encyclopedias that contain more information than could ever be printed, the documentaries on niche topics, and the discussions among world-renowned experts. Of course, it's important to avoid getting addicted to it, but that applies to everything in life. A kid who only reads books and never goes out has just as much of a problem as their classmate who's online all the time.

> It's easy to denounce technology, but sometimes we are unfair in doing so.

Young people may not always delve as deeply into these topics as we'd like them to, but that's not something to be ashamed of. Maybe a child is less interested in math and more into gaming, but this often involves a lot more than just mindlessly playing some game.

You need fine motor skills and have to be able to analyse situations quickly. You must also often be able to communicate in English and have the skills to work together as a team. That's why tech companies like to hire gamers. Who knows, that child may discover how nice it is to put different people together and manage them while playing games, and will take on a managerial role in a company later in life.

While on the dyke one day, I saw two elderly people on a bench playing on their smartphones. It was tempting to joke about how they weren't looking at the sea in front of them or that they didn't chat for a while, but it was a strangely pretty sight as well. Loneliness among the elderly is high, but thanks to digital tools, they can more easily keep in touch with their families or acquaintances. It's easy to denounce technology, but sometimes we are unfair in doing so.

We have to look beyond the idea of someone 'just' staring at their mobile phone. In his essay *Look Down*, Dutch author Daan Windhorst described smartphones as 'a magical window into the world', which I think is a beautiful description. Maybe the two elderly people were indeed watching a cat video or playing some silly mobile game, but they could just as well have been talking to their grandchild in Australia or sharing an article on Reddit about hummingbirds.

It may have seemed to an outsider as if they were cut off from 'the real world', but that distinction has long since disappeared. The technologies Digitalis is built on are directly impacting the world in which homo digitalis lives.

3

The world the homo digitalis inhabits

Is digitalisation disrupting our society?

The changes in Digitalis are impacting the entire world. Innovations in the fields like mobility, housing, energy production, newsgathering, ecology, and safety are transforming our surroundings. Due to new technologies, the only constant factor in human society is change. But that society is more than the sum of its individuals. Think of culture in the broadest sense of the word or of inclusive democracy, which takes everyone into consideration.

← Technology has an impact not only on us but also on our world.

3.1 A cultural revolution

Are we becoming slaves to digital technology?

Digital technologies are impacting our culture, but what does that word actually mean? When we talk about 'culture', everyone immediately has a different image that comes to mind. For some, it's the way we treat each other; for others, it might be statues in the park or old literary texts. The concept itself is somewhat elusive and difficult to define. If you were to ask me, 'culture' is everything humankind creates and therefore doesn't fall under 'nature'.

No matter how you define it, one thing is certain: we're in the midst of a cultural revolution driven by technological breakthroughs. The term 'cultural revolution' has been tarnished by Chinese communists, but humanity has indeed experienced three major cultural revolutions in the past millennia that have always led to significant shifts.

The first was the introduction of the written word in about 5000 BCE. This revolution ensured that knowledge could be passed on much more easily because it no longer had to be done solely through word of mouth. In order to distribute these writings, you still needed a monastery full of monks to make all of the copies, so the spread of such hand-written knowledge was limited. After this came the industrial printing press, which allowed the same knowledge to be spread much faster. In turn, this gave rise to media and allowed for more people to get an education, resulting in fantastic progress for both people and culture.

The third cultural revolution was the rollout of the internet that we are now experiencing: a global network where more than four billion people communicate with each other directly and in real time. The scale at which this is happening is gigantic, and because the filters and borders between people, companies, and organisations are disappearing, the impact on culture in its broadest sense is also quite massive. Culture is already a lively and vibrant thing in and of its own, but because people are now connected to so many others, that revolution is moving faster than ever before. As a result, some people are also afraid that the individuality of their culture may be affected.

This is not necessarily a problem: technology has always caused our language and values to evolve. With the introduction of the bicycle, we adopted the customs of people from other villages; then later, the car and the plane quite literally made our worlds a lot bigger.

What remains important is that we're aware of this and take matters into our own hands. If we do nothing, we risk being overwhelmed as we adopt more and more elements from other cultures because we don't find inspiration in our own European story. But we also shouldn't define that story as something in opposition to other cultures: being proud of where you come from is perfectly possible without being against someone else.

Global citizen and locally connected.

Nowadays you can feel like a citizen of the world but at the same time find pride and comfort in your local community. You can live in a small village and still communicate online every day with people who live thousands of miles away. You are able to be a citizen of a small village, a member of an online community, and a European citizen of Digitalis committed to specific values all at once.

Fewer notifications, more free time

Culture is also primarily determined by what we do outside work, like going to a bar with friends or watching a movie with the family. Some argue that digitalisation is helping those distinctions to disappear because we're spending more and more time on our smartphones.

If you take a look at Google Trends, you can find out how popular a particular search term has been over the years. An interesting example of this is the search term 'social media detox', which increases in popularity every year. There are always moments this term peaks, usually around January and summer break, the months when we start questioning everything. We want to improve our lives by quitting technology, and not just social media, but also smartphones, computers, and apps.

Sometimes the internet is like a bakery filled with pastries and delicious whipped cream. There are so many appealing things that you can be overwhelmed. But like in a bakery, you don't have to buy and eat all the pastries every day, but you do need your daily bread.

Articles appear all the time complaining about how young people are going on vacation with family but don't enjoy the moment and are instead only busy with their online friends. The world never seems to stand still anymore because network signals can send us notifications even on the highest mountains. The feeling of having to be online all the time is stressing us out. Because the human brain already has to make tens of thousands of decisions a day, these extra digital stimuli can quickly feel like too much. Also, when we look at digitalisation's impact on our free time, we often focus on the negative aspects.

The solution is to choose to deal with technology differently yourself. Just because something is possible doesn't mean you have to do it or interact with it. This also applies, for example, to parents who want to Skype with their child every day when they're studying abroad or to managers who constantly send their employees questions after working hours.

The flow of notifications will decrease as the technology improves. Now they're still sent in a quite rudimentary way and it's often all or nothing. By applying algorithms to this problem, it will become easier for technology to determine which notifications you want or need and when you need to receive them. If your smartphone can figure out that you're abroad on vacation and are no longer checking your email, you might automatically stop receiving notifications about meetings. But those algorithms will also need to realise that it's still very important to send a notification when your smartwatch measures a troponin value that indicates a very high chance of a heart attack. You may have no idea what troponins are, but that's okay since your smartwatch will be able to monitor them and warn you when something is wrong.

And this is not the only technological development in that field that will improve for homo digitalis. Ambient computing will ensure that technology isn't as in your face as it is now. Smartphones as devices will gradually disappear and many decisions will be made automatically if we allow the technology to do it for us.

Throwing away your devices is therefore not a good idea. Such an over-the-top reaction is throwing out the baby with the bathwater. Even on a desert island, you'd still want to access the useful aspects of technology: GPS, a source of knowledge, weather forecasts, health monitoring, etc. The idea that you have to fully disconnect to be happy and relax is wrong.

What's more, technology has always ensured that we have just that little bit of extra free time – it was only when mechanisation got underway that the concept of a weekend came into existence. Teleworking saves us a lot of time by eliminating the daily commute. In the past, the free time gained was often filled with things that didn't really help us move forward. In one interview, French philosopher Michel Serres said, 'The 3 hours and 37 minutes we've gained each day thanks to industrialisation are being spent in front of the television, which is just making us... dumber.' We can make better use of the extra hours that Digitalis provides us rather than just spending them in front of the TV.

Thanks to the digital world, we will have more freedom in the natural world, a boundary that's no longer as strictly defined as it once used to be. How many couples nowadays first meet online? And how many random encounters have later turned into virtual friendships?

Pioneer Proust

In the following chapters, I'll focus on how culture with a lowercase *c* (what is not 'nature'), but also that other Culture (opera, painting, movies, etc.) that we consume in our spare time, will change. Creatives and art lovers find each other online, lowering the threshold for visiting famous artistic places. During the pandemic, the Vienna State Opera posted recordings of operas online that could be viewed worldwide for free. Snobs might say that you should see an opera in person to really be able to enjoy it, but snobs always say such things. French writer Marcel Proust had a so-called *théâtrophone* in his bedroom so he could listen to opera performances from his bed. If the internet is allowing more people to come into contact with opera, theatre, and other art forms, it's nothing short of a blessing!

The fear that this lowered threshold will lead to mediocrity is, in my opinion, also unfounded. Earlier I talked about the internet as an endless web of in-depth knowledge.

From now on, the world of art and culture will also be able to find a niche audience for every performance – a theatre company could suddenly have the whole world as its audience and wouldn't have to limit itself to only the local audience.

The fact that people on the other side of the world can follow theatre productions is, of course, due to the language that billions of people understand: not Esperanto, but English. It's important to note that this doesn't mean other languages such as Finnish, Romanian, or Dutch have to disappear. It's important to keep using both one's native language and the global language of English. However, native languages will need to continue evolving; otherwise they'll be at risk of disappearing. But what's most important is that, for the first time in history, we have a language that almost everyone understands, mainly thanks to the internet. This is important, because it helps avoid misunderstandings and makes Digitalis truly inclusive, because Digitalis is there for everyone.

3.2 Digitalis for everyone

Does digital technology offer equal opportunities to all?

It's crazy to think how those with power didn't listen to certain voices for so long. Women, for example, were only given the right to vote a century ago! This has changed thanks to progressive insights and protest movements, but there is still a lot of work to be done.

Fortunately, Digitalis is accelerating everything: movements such as #MeToo and Black Lives Matter started online and have triggered change. And that will happen more often in the coming years because the world in which homo digitalis lives could become a lot more equitable.

I consciously use 'equitable' here and not 'equal' because there are distinct differences in what those terms entail. Fortunately, we don't all have to be exactly the same. Still, we do have to ensure that we give better opportunities to those who are currently disadvantaged, allowing everyone equal opportunities – as much as possible. Imagine there is a fence somewhere that people want to look over. You wouldn't give everyone a ladder of the same height. Children would get a higher ladder than adults. Some people may even be so tall that they can see over the fence without any steps at all, so why should they also be given a ladder based on the principles of equality?

The same applies to Digitalis: some students will need a free laptop or an internet connection subsidised by the government while others won't. If every student were given a laptop anyway, the inequalities that already exist in society would just remain in place.

Equity doesn't mean 'giving everyone the same thing'.

Debunking the disadvantages

Inclusion is a topic close to my own heart because ignoring it can have far-reaching consequences. The discussion about the impact of technology and what our future should look like is still being held solely by a group that's far too homogeneous. The time has come for sociologists, anthropologists, and psychologists to actively participate in the public debate that will interpret and help shape technological evolution.

If we do not take equity into account, we'll end up still living in a class society in the future, one in which existing differences will be exacerbated, which in turn will lead to populism, and we all know how that ends.

Some people will say that the loudest voices are already hijacking the public debate and this is true to some extent. But every era has had its charlatan. Even in the Middle Ages, some people were proclaiming that the end of times was near. I am therefore convinced that wider access to information and a good basic education will debunk such fake news more quickly because people will be less likely to believe whatever they are told. However, when education is lacking, they could remain oblivious to the obvious.

On those same social media platforms, minorities and people who otherwise wouldn't be heard are finally claiming their voice. Black Lives Matter and #MeToo got their start on Twitter, Facebook, and Instagram. We might complain about filter bubbles, and of course they do exist, but these have always been a part of human society. In the past, information sharing was compartmentalised based on which newspaper you took. People who grew up in a Roman Catholic family would read a Catholic newspaper and those who grew up in a socialist family would read a socialist newspaper. The newspapers were always clear about their bias. In the next chapter, I will go into more detail about these filter bubbles, but I would like to say that, thanks to social media, it has become much easier to get in touch with someone who has completely different beliefs. Do you really think your grandparents came into contact with as many diverse voices as you do now?

Others point out that digital developments are causing income inequality. People like Bill Gates have made tens of billions thanks to technological innovations and developers have salaries that most people can only dream of. Again, these inequalities have always existed. Just think of Tutankhamun or the highest-valued company ever: the Dutch East India Company, an Amsterdam company that was co-founded by Dirck van Os from Antwerp. At its peak in 1637, the company was worth 78 million guilders. Taking inflation into account, that would now be about $7 trillion. You could easily buy Facebook, Apple, Alphabet, Amazon, and Netflix for that amount. There are similar companies today: Saudi Arabia's state oil company is worth more than us$2 trillion.

The difference? Unlike Tutankhamun and the oil sheiks, Larry Page and Bill Gates became rich without being born pharaohs or on land with lots of oil and without making others poorer. As I wrote in the introduction of this book, we have all in fact become richer thanks to their inventions: our smartphones alone are worth more than $30 million and we no longer have to buy expensive encyclopedias to gain knowledge.

We're only halfway there

Making Digitalis an inclusive and equitable place starts with ensuring that everyone has access to the internet. Currently, more than four billion people are connected, which is great, but that also means that about three billion people still aren't.

This is probably a conscious decision for some people, but many others are involuntarily cut off from Digitalis. The first step is to make sure that connections and mobile coverage are possible everywhere, which is the responsibility of the telecom companies. However, it's not always profitable for them to provide service to specific areas. In this case, it becomes the shared responsibility of those private companies and the government to find solutions. Access to the internet is not yet a basic human right such as security or privacy, but this must change quickly because the internet is becoming increasingly necessary to function in society.

Having access is one thing, but knowing how everything works is another. Some people have little intrinsic motivation to work with technology, but certain situations can change that. Just as companies rapidly developed their e-commerce offerings and switched to teleworking during the Covid shutdowns, a large group of people discovered the benefits of technological tools in just a matter of weeks. People started using video calling en masse in residential care centres because it was virtually the only way to see family and friends in the outside world. Residents quickly learned how to talk to their loved ones with the touch of a button, although they may never have used a PC in their life.

When people are open to such innovations, they discover that they can often make them more independent. Hopefully, self-driving cars will soon ensure that people who are less mobile will be able to get around more easily in the future. In the same way, VR glasses will be used to allow older people to 'travel' again and classes will be able to 'visit' historical places instead of just reading about them. This is one of the ways technology will be used to make the world genuinely inclusive.

Technology will continue to fulfil that role, allowing us to welcome more and more people to Digitalis. The people developing digital applications will need to take this into account because not every user is the same. The movement around inclusive design is thus rightly growing. This way of working means that developers will pay more attention to users with specific needs. For example, it's hard for visually impaired people to visit unstructured websites, because their screen readers can't convert the text into sound. Software used for video calling is often not adapted to the hearing impaired because the person speaking is not fully visible – which is difficult to register in the case of sign language, but live captation thanks to AI will help.

By involving those users in the development process of new applications, such errors can be avoided. Hardware manufacturers should also focus on this. Just like one can use a telephone in both Spanish and French, they need to consider the visually impaired, hearing impaired, and others for whom technology is less obvious but can provide tremendous added value. For example, some applications make it possible to add a Braille display to your smartphone, and blind people can have screen readers to read everything on a screen. With Google Lens, you can now scan a text and have it read or translated so that people who can't read or speak the language can still understand the world around them. Anyone looking at it from a pure profit perspective would ignore such things, but anyone who believes that Digitalis should really be inclusive will make sure this happens.

Made for and by a diverse society

When entrepreneurs explain in interviews why they started their business, they sometimes say that they had encountered a particular problem for which there was not yet any solution. To fill that gap in the market, they started a company, with the first employees usually being friends or fellow students, people they had a connection with – in short, people like them. After all, they'd likely be spending more than 40 hours a week together.

That's an entirely human choice that can have far-reaching consequences. These first employees, in turn,

We need to build a diverse, inclusive, and equitable world.

mainly recruit colleagues who are similar to them and sometimes even receive a bonus if they introduce a new colleague from their circle of friends. This can quickly tilt the orientation of the company in one direction. This way, a chain reaction is set into motion, because when young people see that there is only one type of person working in a particular sector, they become more or less inclined to start working there themselves.

That's precisely what has happened in the tech world in recent years. Initially, many women were employed in that world. Until the middle of the last century, mathematics and computer programming were mainly tasks done by women. Yet pioneers like Ada Lovelace and Grace Hopper could not keep the technological sector from gradually becoming a white man's bastion. It's worse in some niches than in others: game company Ubisoft, for example, has 18,000 employees worldwide, but only 22% of them are women. At the head office in France, it's just 14%.

If a diverse group of people is not developing the technology, it can have serious consequences. Just look at what has happened to the development of artificial intelligence in recent years. Artificial intelligence is a technology capable of performing human tasks, such as reading a text aloud or interpreting an image. The latter is called computer vision and is used for such things as licence plate recognition.

> If a diverse group of people is not developing the technology, it can have serious consequences.

The algorithms behind that artificial intelligence are able to do this because they've seen a lot of photos of a certain object, after which they eventually start to recognise patterns.

This works quite well, as long as the data used to train the algorithms is good. Suppose, however, that the algorithm is consistently being told that photos of a dog are actually pictures of a cat because the images are incorrectly labelled. In that case, the algorithm will, of course, no longer be able to recognise a dog. Developers often jokingly say, 'Garbage in, garbage out.'

The dog example is pretty harmless, but things can quickly go wrong when using algorithms to make important decisions. For example, various police services in the United States use such algorithms to track down criminals on the street using smart cameras, among other things. Because those algorithms often train on photos with mainly white people, they're significantly more likely to fail when they have to recognise a black person. Several innocent citizens have already been arrested because the algorithm wrongly confused them with a wanted criminal.

When the Black Lives Matter protests broke out, tech companies like Amazon and ibm decided that their software should no longer be used for such practices, but in reality, those mistakes could and should have been prevented. A study published by the us National Institute of Standards and Technology in 2011 showed that similar algorithms developed in China, Japan, and South Korea are much better at recognising Asians than white people. It therefore only makes sense that an algorithm developed by an all-white group of engineers would have trouble recognising black people.

Yet, there is no reason to panic because artificial intelligence is not in charge of our society. Humans are still the ones making decisions. Nonetheless, a legal framework is needed and a diverse group of people must develop ai. In addition, as mentioned earlier, the data on which the algorithms are trained must also be as neutral as possible.

This mix of people will also ensure that the company benefits: research by the World Economic Forum shows that diversity in the workplace stimulates innovation. BCG found out that there is a very strong correlation between the diversity of the management and the business results: diverse companies are able to better understand their diverse customers and the diverse world they operate in, so they are more successful.

This diversity will not come about by itself: employers need to actively seek out such candidates. Instead of hiring candidates that match the current company culture *('match culture')*, the question is which candidates will add new elements of culture to the organisation *('add culture')*.

Diversity in the workplace is achieved not just through targeted recruitment but mainly by creating an inclusive corporate culture. This won't happen on its own either: it's an active process that consists of different phases. There are many steps between recognising racism or sexism in the workplace, whether directly or indirectly, and openly and actively combating it.

Diverse companies are able to better understand their diverse customers and the diverse world they operate in, so they are more successful.

It was only when I started working at Google that I experienced firsthand that mixed teams are usually more successful. Matters are questioned more quickly because everyone brings different perspectives. Therefore, it's not only about diversity but also about how everyone can subsequently have their say and have their opinions taken into account. Like many others, I have also experienced the power of working together in a rather bizarre form of entertainment called the escape room.

I once took a trip to Silicon Valley with some business leaders and their children. We all went to an escape room together and the group was divided into two: the teenagers and their parents. The first group solved all the puzzles and left the escape room in less than half an hour. Our group eventually had to be liberated by the organisation after more than an hour. We hadn't been able to work together because, all being macho men, we reasoned too much for ourselves without sharing our findings out loud and working together. It immediately seemed to me to be a perfect metaphor for society in which everyone is all too often working on their own little island. It's only by working together, sharing findings, and showing genuine interest in everyone's proposal that we can solve complex problems.

There are many steps between recognising racism or sexism in the workplace, whether directly or indirectly, and openly and actively combating it.

3.3 Coexisting in Digitalis

Do social media and fake news lead to populism in Digitalis?

Humans have always clung to tradition. We have a strong desire for certainties. We want to be able to explain even the inexplicable and therefore sometimes resort to what in actuality are blatant lies.

Uncertainty is what feeds the latter the most. The people taking advantage of those uncertain times to deliberately spread false news and rumours know this too. They say that the Covid crisis is a conspiracy of the rich or that Bill Gates is sponsoring vaccinations so he can inject microchips into people. Sometimes these messages even come from major media sources, such as Fox News or RT in Russia. Hoping to create order out of what is an unpredictable situation, people all too readily start to believe these 'alternative truths'. Since these theories often offer a clear, if seemingly miraculous, answer to a complex problem, they are rather successful at catching on.

From fantasy to propaganda
There are a lot of messages online that are not entirely correct. Usually, they are quite harmless, but sometimes they are downright dangerous. Facebook posts that are slightly too enthusiastic or an Instagram photo that has been spiced

up with a filter may be no big deal, but fake news stories are a real problem. These may have been written by someone who wants to push their own opinions by deliberately concealing certain things or even lying.

In the best-case scenario, these stories come from a fantasist who gets a kick out of how many people interact with their post or is just a troll wanting to have some fun. It gets a little more serious when people build business models on them. According to the Global Disinformation Index, a British non-profit organisation that studies disinformation, fake news websites earn more than $200 million a year.

The most dangerous scenario is where such false messages are used for political purposes.

The most dangerous scenario is where such false messages are used for political purposes and to make people behave differently. Of course, the most notorious example of this is Cambridge Analytica, a data company that sent very targeted messages (whether false or not) to Facebook users based on psychological research.

Social media's fault? Only partially

The fact that extremist parties are responding to specific fears is nothing new, nor is the intentional manipulation of photos and videos. Sometimes they go too far, though, and the manipulation is very obvious. For example, in the USSR, Stalin cut away political opponents like Trotsky and

Kamenev from historical photos in an attempt to change history.

The big difference with the past is, of course, that the techniques for doing such things have become much more generally accessible. Where Stalin's henchmen still had to work with scissors in the darkroom, this can now be done with just a few clicks of the mouse. Those technologies also allow you to create so-called 'deepfakes': manipulated videos in which someone appears to be saying or doing something that they haven't actually said or done at all.

Social media is quickly being blamed when elections yield unexpected results. For example, the outcome of the Brexit referendum, in which millions of British people voted to leave the EU, was, according to many, due to false news from the leavers camp.

I don't entirely agree with that statement. Brexit is not primarily due to social media but mainly to the tabloids, which have set the tone of the debate in the UK for decades. The British tabloids, in particular, have had a knack for blaming everything that went wrong in the past fifty years on the European Union. However, the great danger in all of this is that you no longer have to wait fifty years to influence a population.

In addition to the rise of fake news, social media has also hardened the public debate. Increasingly, the world seems to be divided into two camps and harsh words are being used online.

What should we do?

There are several things we can do to gain control. On an individual level, we need to make people more critical of the media and teach them how to engage in online debates in a civilised way. As I mentioned before, education plays an essential role in this, but people also have to remain alert as they get older.

Traditional media also has a lot of work to do. Journalism must be able to prove its quality and not amplify malicious messages. By treating some Tweets from extremist politicians as news items without context, they amplify the attention given to those messages. Twitter mainly proves that anyone can be an 'expert' nowadays. Things are deliberately posted on there to be picked up by the media.

We need to make people more critical of the media and teach them how to engage in online debates in a civilised way.

News organisations could, for example, join forces to develop a label for quality news. A logo placed on videos or with newspaper articles could indicate that they're from a reliable news source. This strategy could work since young people can indeed distinguish an unverified Tweet from a newspaper article. Strict action must then be taken against sites that abuse this quality mark. By working together on fact-checking, editors can fight this misinformation.

In addition, journalists must also indicate their sources more clearly. When a newspaper reports that research shows electric cars allegedly pollute more than diesel cars but doesn't mention that this research was funded by the car industry, it doesn't help the audience get an objective picture.

Advertisers also play a major role in this because they don't want their advertising to appear next to fake news or hateful messages. Websites like Sleeping Giants alert major advertisers when this happens and those brands will blacklist the sites in question. They also force big tech companies to take strict action against online hate, and that's a good thing, although those tech companies shouldn't have waited to be scolded before taking action.

This way, they can ensure that fake news and hateful messages get blocked, but even more important is putting forward reliable news sources. Google has already had its own digital news initiative for a number of years. It provides financial support to media to stimulate innovation, such as combating fake news or hate messages. For example, in Asia, the fund has contributed to developing a tool that can recognise fake images and remove certain photos from the internet.

Tech companies such as Twitter are, however, not a court of law. They can delete specific Tweets or place fact checks next to blatant lies, but often an infringing party can just create a new account and carry on.

The question is also whether we can leave it up to tech companies to make this judgment because these are discussions that are fundamental to preserving democracy. Take the 'right to be forgotten', for example. Citizens of the European Union can contact a company and ask them to remove outdated or incorrect information about them. This mainly concerns search engines such as Google, where people can ask that certain links mentioning them be removed. But should it really be up to a search engine to determine what may or may not be removed? Google only refers users to other sites: removing the link on Google doesn't remove the original reference from the internet, just like a malicious entrepreneur doesn't disappear just because their company no longer has a business licence. Google is forced to act as a judge in these situations, but in fact, the (news) sites are a much better place to look at when discussing the balance between the right to information and free speech and the right to privacy.

It also isn't always easy for a court of law to determine the rules in Digitalis. To penalise hateful messages, you need to be able to trace who posted them, but that's easier said than done. Forcing everyone to post everything under their real name is also dangerous at times – in an undemocratic country, online anonymity is vital for some minority groups. Nevertheless, in Europe, we could make it compulsory to log in (e.g. in Belgium with *Itsme*) some news sites in order to respond to or trace official messages from politicians.

An important principle is that there should not be a difference between online and offline: the same rights and responsibilities should be respected online as have been in the past. In most countries, you can't insult a politician in the street, so why is it still possible online?

Populism: a constant threat

If there's one thing you have to remember, it's that the greatest danger isn't technology but humans. Populism has always been a threat. After all, the Nazis came to power without social media. Thanks to his minister of propaganda, Joseph Goebbels, Hitler soon realised that the spoken word worked better than written texts in his case and therefore decided to distribute cheap radios to the people. For every medium, there has been a populist who abused it, and online media won't be an exception.

The greatest danger isn't technology but humans.

Still, there's no need to be too suspicious: if it were that easy to influence people through technology, we would have had at least ten Nazi states in twenty-five years of the internet. We're still dependent on what people do with technology, but it's important to recognise that democracy as we know it is not a given. We have had to go through some very dark times to appreciate its importance, and we are always at risk of reaching such a low point again. But democracy can also get better.

3.4 The world is not yet done for

Isn't technology only aggravating climate and environmental issues?

Our earth is 4.5 billion years old, but in fewer than two hundred years, we've managed to pillage its resources at a frantic pace. We have cut down trees, mined precious metals, drilled for oil and gas, and so on. We did this all in the name of progress – and progress we've made. Fewer people live in poverty, child mortality has been significantly reduced, and people are living longer than ever before. That's fantastic news, of course, because we have never had it as good as we do now. You read that right: we've never had it this good. Never doubt this again! Feel free to refer to a history book or dataset to substantiate that claim.

It is nonetheless inappropriate for rich Westerners to prohibit others from enjoying the things that many people have here, but it's also impossible to continue consuming the world's resources at this rate. We thus have a huge problem because this is far from the end of the marathon that we're all running together. We'd all like life to continue on this planet for a few more billion years if at all possible.

The solution isn't to use less, but instead to create more technology to solve the major problems we have in the world.

Digitalisation offers solutions for a greener society.

The solution isn't to use less, but instead to create more technology to solve the major problems we have in the world. The old system needs to be overhauled so we have the opportunity to use technology to develop a sustainable way of life that benefits everyone. A society that doesn't run on combustion engines or natural gas but revolves around people and their environment – one where we might even manage to have a positive impact on the climate.

Technology will help us get more energy

The previous Industrial Revolution relied heavily on finite resources such as coal, oil, and natural gas. Yet the sun alone gives us more energy every day than we'd need for an entire year – we just need to be able to collect and store it. We have now reached the point where this can be done on a large scale and in a profitable way. In addition, we can now also extract energy from wind, ocean waves, and the earth.

When Europeans realised that installing solar panels had a direct impact on their energy bills and were given tax incentives to install them on their roofs, many switched. Meanwhile, the technology has become so efficient that after about ten years the energy is completely free – my bill has since been reduced to almost zero. Of course, it remains a significant investment, but companies such as Elon Musk's SolarCity also offer users the option to rent solar panels. This could already be enough to help make the technology more democratic.

In recent decades, there have been many technological breakthroughs that have improved the ability to collect and store that energy. The efficiency of solar panels has also increased considerably, making it more affordable to install them yourself. Thanks to home batteries from Tesla or the German VoltStorage, the energy from those solar panels can also easily be stored. Hydrogen is now also being used for auxiliary or main engines on freight ships.

But is it sufficient? The European target is to get 32% of energy production from renewable sources by 2030. In Copenhagen, the government has already raised the bar

The raw material of the fourth Industrial Revolution is not oil, but knowledge.

considerably: by 2025, it hopes to become the first climate-neutral capital in the world. We should prioritise such ambitions in more regions.

Google has gone even further. Despite the significant need for electricity to run its data centres around the world, the company has been carbon neutral since 2007. As of 2017, all of the energy Google consumes has come entirely from renewable sources. In 2020, the company even compensated for its total CO_2 emissions since its inception 22 years earlier. Hopefully, this will serve as a source of inspiration for other companies, organisations, and governments.

As of 2017, all of the energy Google consumes has come entirely from renewable sources.

We therefore can and must invest more in green energy. Our governments are investing billions into researching new, still uncertain technologies such as nuclear fusion, while there are already working technologies available that can offer a solution.

We don't just talk about energy when we talk about raw materials. The shortage of potable water is also becoming a major challenge, but by combining that abundance of energy with new technologies, we could successfully convert the endless supply of sea water into drinking water. There are desalination plants in Israel, which are becoming less polluting due to the use of solar energy. This is fantastic news and a huge leap forward from an ecological point of view.

The next step includes the technologies that will ensure that we can also provide all residents of Digitalis with quality food without further endangering the environment. Some devices already offer a glimpse of what may come. For example, Urban Crop has developed a modular system for doing vertical farming in a smart way. This invention allows vegetables to be grown almost anywhere, with no need for pesticides and with much less water wastage.

By using soy milk instead of cow milk, we disrupt in fact the whole -very polluting- process of the cow. We can go from vegetables to milk directly, without animal suffering.

Even for those who can't completely leave meat out of their diets, technology offers a solution, thanks to breakthroughs in imitation meat and in vitro meat. We have now reached the point where cultured meat is cheaper – and definitely more environmentally friendly – than animal meat.

Technology will help us be more economical with raw materials

Technology will also help keep us from wasting raw materials. The best way to illustrate exactly how this will work is with an example:

For years, Philips grew with the development of products such as light bulbs. Bulbs were something that you had to replace every year, so the company was sure of a steady source of income. That changed in the early 2000s when the much more sustainable and smart LED bulbs emerged. The company had to reinvent itself, changed the name of its lighting branch to Signify, and, to many others, became a service company. From then on, the company was no longer selling light bulbs, it was selling light.

It is a similar transformation to the one we've seen in other sectors and which has resulted in a shift from 'owning' to 'using'. Homo digitalis no longer buys music on discs in plastic boxes but instead pays for a subscription to a streaming service such as Spotify. We aren't as likely to own a car anymore, but instead use shared vehicles via Poppy or Zipcar, for example. Companies no longer buy light bulbs, but instead are customers of Signify, which ensures that there is sufficient light in the office via a subscription scheme. There will always be people who prefer to own their own, often more expensive, models of things, but a growing group of people is finding this less important.

This, of course, impacts the production of devices because, from now on, it's the manufacturers who will be the guarantor when something breaks. It will be the final blow to 'planned obsolescence', a cunning method of deliberately giving products a limited lifespan. Imagine how sturdy a pair of tights would have to be if manufacturers had to send you a new pair as soon as a tear formed! The products used by homo digitalis may no longer be their own property, but they will last much longer.

By digitalising things like lighting, you can also optimise them. Because things are measured, you can compare different situations using machine learning. Signify does this already. By using data analysis, the company estimates how much light is needed in offices or football stadiums as accurately as possible. Parameters that play a role include the amount of sunlight and the number of people present.

Similarly, self-driving cars can now use less fuel and Google Nest thermostats can determine the ideal temperature in your home and stop heating when you go out. These are good steps towards more sustainable consumption, but still small potatoes compared with what's possible on a larger scale.

LED lighting may be more durable than old-fashioned light bulbs, and we can increase the impact by using them with care. Illuminated public areas are growing worldwide by 2% a year, but smart street lighting can offer a solution.

These days in Belgium, we illuminate large parts of the entire highway, regardless of the number of cars on the road. In the future, we will reverse that. There will always be a dim light, but when someone approaches, the lights will start to shine brighter a few hundred metres in front of the car. This will also be a lot safer for cyclists and pedestrians, who will be able to see that a vehicle is approaching well before the headlights come into sight. The smart lampposts can make constant adjustments based on data and collect information about traffic on the road.

These smart devices not only will affect how much energy we'll save, but could also save your life – something I've experienced myself. You wouldn't expect it from me, but I'm usually not the first to adopt the latest gadgets. When smart Nest Protect smoke detectors were on sale around Christmas, I decided to buy them because I needed new smoke detectors anyway.

Several days later, one of the smoke detectors gave a signal, and I thought I had made a mistake when installing them. To my great surprise, a voice told me there was a carbon monoxide alarm going off in the basement. A simple smoke detector would have only beeped, which I wouldn't have heard on the second floor of my house. The Nest unit could tell me where something was wrong and I could quickly go downstairs to fix the problem.

Onwards to less waste and a fairer system

We now throw away a lot of food, but digitalisation can ensure that we will waste less. A customer standing in front of an empty yogurt shelf in one supermarket doesn't know that, a few hundred metres away, another supermarket has one that's almost out of date.

By gaining more insight into what, where, and when customers buy things, supermarkets can better tailor their stock to their clientele. At the moment, this is often done based on gut feeling. For example, supermarkets are full of school supplies at the end of summer, but this could be timed much more precisely.

The next step is for homo digitalis to directly contact the producer, as is now already happening little by little via online platforms supporting local farming. The distance that the products on that kind of platform travel from field to plate is ten miles on average, which is a lot better for the environment than having apples flown in from Spain.

Consumers who buy avocados or kiwis in the supermarket often don't realise that their food is flown in from all over the world. Just as with the Nutri-Score, an index that shows how healthy a product is, there could be a 'transport score' or a 'sustainability score' for our food.

This doesn't have to be limited to the food industry either. From the 1960s onwards, the fashion world began to massively relocate its production to low-wage countries.

Thanks to robotisation, it will become almost as cheap to produce clothing in our own country again, which would, of course, be much more ecological because the transport costs would largely disappear.

This will also make it possible to respond more quickly and even better to fashion trends since weeks often go by between production on the other side of the world and sales in Europe. This way, less unsold clothing would be thrown away because the supply could be better adjusted to the demand.

Such a transport score could also be used in many other ways, such as determining how much producers have to pay in taxes. Those who want to produce something in their own country in a fair way currently often have to pay a lot more in taxes than those who go for the cheapest workers, despite the ecological impact. By including factors such as transport, pollution and paid taxes, we could end up with a much fairer system.

Is the technology sector itself even that sustainable?

You could, of course, argue that the technology sector itself consumes an enormous amount of energy.

500 hours of videos are posted on YouTube every minute. 6,000 Tweets are shared and two million emails are sent every second. There are over a billion websites, and by 2020, we will be creating 35 zettabytes of data annually. That's 35,000,000,000,000 GB. These are staggering numbers that we can hardly imagine.

You might be thinking, 'So what?' What difference does it make if you add a few ones and zeros? It's an extra line of code; it doesn't bother anyone, right? Not quite. What at first seems relatively unimportant can have a huge impact on the survival of our planet.

As nice as the name may sound, the data that we store online doesn't end up in a 'digital cloud' at all. It's stored in data centres, which you could safely call the brains of the internet. They are often big buildings, with few people inside. For example, Google has large data centres in Belgium, Finland, Denmark, the Netherlands, and Ireland. Thousands of servers 'live' in these bunkers, ensuring that the internet continues to run.

The telecom and data industries represent a solid 10% of the world's electricity consumption. Indirectly, a smartphone today consumes more energy than a refrigerator. Not because we charge it so often, but because the data centres (and other intermediate links) consume that much.

One of these server parks consumes about as much energy as your average city. The exact same thing happens in data centres that happens when charging your phone or computer or when you are trying to run several tasks at the same time: things get very hot. So hot that the room has to be continuously cooled with air conditioning for everything to keep functioning.

Fortunately, the sector has seen rapid improvements in recent years. The same machine-learning algorithms from DeepMind that defeated the world champion of Go in 2016 were used by Google to explore how to maximise data centre efficiency. The company is now saving over 30% in energy. It would never have been possible for a person to determine so accurately how to optimally cool a data centre. By using only

Data centres are becoming more energy efficient.

renewable energy, Google has been able to be completely climate neutral since 2017. Apart from those efforts, the digital sector has already made it possible to dematerialise products like encyclopedias, film rolls, newspapers, video cassettes, etc. that don't have to be produced and transported anymore: the positive ecological impact of this dematerialisation is almost immeasurable!

The next step? Climate-positive companies that will remove more greenhouse gasses from the atmosphere than they emit.

Now that the energy consumption of the data centres is being decarbonised, the devices we use have become the biggest problem in our industry. They are made with very precious raw materials, such as gold, silver, and palladium, the mining of which is often done under dangerous conditions. However, these precious metals can be reused. Indeed, there are listed companies such as Umicore that specialise in this.

What's more: batteries of large devices, such as electric cars, can often be given a second life when they can no longer be used in vehicles due to wear and tear or after an accident. In hospitals, for example, they can be used as an emergency battery for several years before finally being recycled.

We therefore need to find a way to ensure that our smartphones and smart speakers last longer and become much easier to recycle. When I was young, there was hardly any recycling. In fact, plastic was hip. Now it's a different story and my children enjoy going to Think Twice or the thrift shop. Just like with our energy problem, we don't have the time to wait until things 'will work themselves out'. We need to take matters into our own hands and require manufacturers to use recycled materials and ensure that their devices are also recyclable. Making appliances more sustainable is sometimes about the little things. The so-called dark mode, which gives websites a black background instead of white, consumes a lot less energy and is better for your concentration.

Meanwhile, in San Francisco, 80% of its waste does not end up in landfills but is instead recycled or turned into compost. The city achieved this by making recycling mandatory and introducing financial incentives that encouraged people to do so. The polluter has to pay while those who do it right get a bonus.

The most important requirement for such a transformation isn't the technology because that is already there. It doesn't have a solution to every problem, but it does offer opportunities. What do we need then? Ambitious politicians, business leaders, and people like you and me who understand this and want to initiate positive change now.

We need ambitious politicians, business leaders, and people like you and me who understand this and want to initiate positive change now.

3.5 A safe life for the homo digitalis

Will Digitalis lead to Big Brother?

Digitalis is a great country, but it has a number of things that aren't entirely under control yet. Just think of privacy, a value that has been at the centre of many debates surrounding technology in recent years.

This debate often includes the term 'Big Brother', made famous thanks to George Orwell's book *1984*. The book was written between 1946 and 1948 and warns against totalitarian regimes. In *1984*, an omniscient political body monitors the actions of all citizens through cameras, both indoors and out. In reality, today's technology is capable of a lot more than just the cameras deployed by Big Brother in *1984*, thanks to the smartphones that we constantly carry with us.

This still goes wrong regularly because the development of new technologies always proceeds in waves. Just look at the early years of electricity. For a long time, heavy appliances such as refrigerators had no electrical grounding. Fortunately, there are now many safety systems in place that ensure we no longer electrocute ourselves quite so easily, which was a regular possibility in the past. French artist Claude François even died from electrocution in 1978 trying to straighten a loose light fixture while standing in his bathtub. Houses used to burn down quite often due to short circuits.

Exactly the same applies to how we now deal with privacy and the online security inextricably linked to it. Houses may no longer be literally burning down, but people's careers and lives are. Just ask a politician how devastating a leaked email can be or consider the impact when a teenager's sexts get forwarded. The fact is that our privacy has seemed under threat in recent years and the answers lay in the hands of three players: companies, regulators, and…ourselves.

Facial recognition has always been around

When we talk about how companies handle our privacy, we immediately think of the Facebooks of the world, large tech companies that can monitor the online behaviour of billions of people every day. Yet it's much broader than that, and it's often the smaller companies that are less careful with online privacy because they don't always have the resources or knowledge to do it properly. A florist that stores a list of customer addresses in an insecure manner, a restaurant owner who writes down with pen and paper who comes to dinner during the Covid pandemic, or the supermarket that doesn't bother changing the default password of the security camera so the whole world can watch what people are doing and buying in the store.

> It's often the smaller companies that are less careful with online privacy.

112

People often approach me after lectures to ask about how tech companies can see everything they buy online, but the combination of shopping and an alleged invasion of privacy is by no means new. Once upon a time, shy people didn't dare to buy condoms from the pharmacist because, before you knew it, the whole town would be aware of your plans for the night. It was even worse for the purchase of AIDS drugs or Viagra. When you buy such drugs online, your privacy is much better protected, if only because strict laws prohibit people from snooping through your purchase history.

Big Brother is watching you.

However, it's only logical that many people are critical of tech giants. Their business model usually revolves around the sale of advertisements, which they can target more effectively by analysing their users' online behaviour. In my opinion, it has been a mistake to make the internet depend almost entirely on ad revenue. That idea was taken from TV, but this ad-supported 'free' model is not sustainable.

> In my opinion, it has been a mistake to make the internet depend almost entirely on ad revenue.

The expectations were wrong from the start because commercial television channels, along with a handful of newspapers and radio stations, had almost exclusive rights to advertisements. As a result, advertising at the time was expensive and therefore very profitable. It's a bit different in Digitalis, where every individual has become their own medium and advertising has become cheaper.

In the end, the world of television has also shifted and we now pay for Netflix, Disney+, and Streamz. You see the same thing happening online on news sites, which are putting up increasingly strict paywalls. The headlines can still be viewed for free, but for real journalism, you have to pay, and rightly so. The same applies to apps and services where an advertising model is less likely to be used. Twitter, for example, is already thinking about a paid version of the platform. That said, we have to defend an open web, funded by (targeted) advertising, so that everyone can continue to access general information and nice blogs for free.

This could reduce the anger people feel about the collection of their data. At the same time, the value of data is lower than perceived: data is abundant and also very volatile. For example, Waze is only interested in where you are now; it doesn't care about your location data from yesterday. Whoever says that 'data is the new oil or gold' is wrong. You can't stash data away and hope to sell it at a higher price a month later.

Eric Schmidt, the former Google CEO, always said that the company would be up and running again after a month if Google would suddenly lose all of its data. The company would only be in trouble if it lost all its engineers in one fell swoop. It's not about the data but the people who develop creative solutions to use it. Technological developments have given us better insight into which data is useful for a specific service to function and which is unnecessary.

Other technological developments will ensure that less data has to be transferred between your device and a general server. That's a lot safer because this limits the chance that the data is intercepted or subsequently stored in a central location. Artificial intelligence can already work autonomously on smartphones, cameras, and other smart devices.

It's not about the data but the people who develop creative solutions to use it.

Thanks to this decentralised way of working, we can develop applications in a very privacy-friendly way. Consider, for example, the contact tracing app governments want to use to trace people who have come into contact with Covid. People immediately thought that a central database would be set up somewhere, containing all the contacts, location data, and appointments, but that wasn't necessary at all. Every smartphone can receive an encrypted code that is exchanged with other devices via Bluetooth as soon as they are less than six feet away from one another. That information is then stored on the device itself for fourteen days. If you later test positive for Covid, a notification will be automatically sent to everyone who has your code on their phone. It's all done completely anonymously, so you won't even know who you potentially infected.

When critics proclaim that such apps aren't allowed because of privacy laws, this shows the unjustified lack of trust in technological solutions. Some countries decided to work with a call centre to do contact tracing, but that's even more intrusive to your privacy. Imagine you're having an affair or you're secretly applying for another job; would you pass on those contacts when a complete stranger called you? The myth of privacy has cost us precious time and money. Six months and $100 million for the region of Flanders only, to be exact. Hopefully, this book will help us avoid such mistakes in the future.

The internet as we know it will always continue to evolve. This is happening under the influence of new regulations, such as the EU General Data Protection Regulation (GDPR), but a lot is also changing on a technical level. For example, we're also looking for suitable methods to additionally

protect the privacy of users. Just think of Tim Berners-Lee's
Solid project.

Privacy above all else?

Thinking that entrepreneurs will suddenly see the light and
develop completely privacy-friendly applications is a bit
naive. That's why there is a need for strict and effective regulations, and with the GDPR, European governments have
shown that they are increasingly committed to doing so.

Nevertheless, it's important not to regard privacy as an exclusive right that stands above all other rights. If we had 100%
privacy, it wouldn't be possible to track down criminals. The
right to privacy must also be measured against the right to information. When a politician commits a significant violation,
the first right is lost in the public interest.

Privacy commissions are often set up as bodies that study
the right to privacy entirely independently of other rights. If
regulators want to work on a good privacy policy, they'll
have to ensure that the various rights and obligations are
better aligned.

A safer internet starts with us

You can't separate privacy from security, and the relationship between the two will only become stricter in the
coming years. For instance, when it comes to hypersensitive

information such as our DNA data, it will need to be stored even more securely than location data.

When sensitive and personal information is leaked, the consequences can be disastrous. Just ask the users of Ashley Madison, the online platform for people looking to commit adultery. When all of its data became public after a major hack in 2015, some of the exposed men ended up taking their own lives.

Your passport to travel safely in Digitalis.

The fault, in this case, was the poor security of the platform, but we individuals also play a role. We must always choose a strong password and use two-step verification. We're also increasingly in control and can decide for ourselves which data we share or not. As a user, you should therefore think about what the value of an online service is for you, before sharing information about yourself. An excellent example of this is Waze. The application needs to know where you are, but in return, it gives you information about traffic jams on your route. People will even consciously share extra information about accidents or speed cameras because they know they can also use the information other people provide. However, it's not always easy to find the balance between sharing information and the added value of an online service, especially when a new app wants immediate access to all your contacts.

The next step is so-called confidential computing. The user data must always be secured when it's stored, moved, or used.

We've arrived in a world in which homo digitalis has insight into what happens with their data and can also actively manage those data flows on a personal dashboard. There is a privacy check under your Google Account, a tool with which users can indicate which activities may and may not be tracked and to what extent services may be personalised. With the click of a button, data can be modified or deleted. Those who just complain that tech companies know everything about them need to put themselves in control of the buttons.

4

Now it's up to us

Are we still in control?

Homo digitalis shouldn't simply be sitting back and letting all of these changes happen. We can bend technology to our will and use it consciously to improve our lives. And we're not alone in this. Digitalisation will impact everything and everyone; all other people, organisations, and governments are facing similar challenges.

> We can bend technology to our will and use it consciously to improve our lives.

The flood of innovations rushing towards us can feel paralysing. You may not know where to start in order to get a grip on the changing world, but it's quite simple.

You have already taken the first step, which is always the most difficult. It all starts with getting information and, hopefully, this book has already helped you with that. It can, may, and sometimes has to rub you the wrong way. I wrote earlier that in turbulent times we sometimes believe in blatant lies because they seem to provide an answer to all of our questions. So always be very critical of what you're told and consciously look for extra knowledge and insights.

The second step is to tame the new technology. This must stem from an intrinsic motivation, so first think carefully about what you want to achieve. Maybe you want more contact with family. Or you want to develop yourself or take the next step in your career. Technology can help you achieve these goals, so you learn to deal with it in a useful way. Technology is a means to an end, not the end itself.

It's up to us to tame and use technology.

Then there is one last step: take action! Perhaps you'll take a course or learn to use a specific tool. Maybe you'll decide to change jobs because the company you work for isn't digitalising enough. Perhaps you already own a business and want to find out which innovative technologies will change your sector. Whatever the case may be, I definitely recommend reading my previous book, *Digitalis*. No matter what your goal is, don't give up, but instead roll up your sleeves and get to work!

Technology is a means to an end, not the end itself.

A new revolution

It's not only up to us as individuals to get to work; policy-makers need to do the same. The Europe of today was shaped by its leading position in previous Industrial Revolutions. The United Kingdom was the first country in the world where the first Industrial Revolution developed and where trains ran.

We live in the midst of a new revolution, yet many politicians sometimes don't quite seem to realise this. Digitalisation, of course, doesn't fall neatly into the portfolio of just one minister. It's not just about technology; it's about humanity. It transcends all ministries and powers.

Digitalisation, environment, and inclusion are problems that humanity needs to solve without the boundaries established in previous eras.

Europe has been at the forefront of prior industrial and cultural revolutions.

Companies and organisations shouldn't just copy their old corporate economy when they digitalise. No, we have to reinvent companies by providing them with a better, more ecological, and more socially responsible way of working.

We can do it
We have to get rid of the idea that other parts of the world are doing better. There's no need for that jealousy because that's not how we will get ahead. We should especially draw inspiration from others because we know Europe can do it. What's more: we can be the vanguard of a new internet,

where business, privacy, and citizen interests can go hand in hand.

The internet as we know it today was invented in Europe. Sir Tim Berners-Lee conceived the concept and founded the World Wide Web in the early 1990s at CERN in Geneva, together with Belgian Robert Cailliau.

We've always been at the centre of the research and we don't have to be so modest about it. This is also a message that we should be able to pass on to the general public: even in Europe, we can dare to dream of innovations that will change the world!

Not only the internet but also a lot of the hardware used in Silicon Valley has been developed in the heart of Europe. This is happening in Leuven, home to IMEC, the largest independent European research centre in the field of micro-electronics and nanotechnology. IMEC works with almost all major tech companies and its technology is in almost every chip used worldwide.

Some companies feel so American that we almost forget they're European. Just think of Spotify, Revolut, or Klarna. Many Europeans simply don't know that those companies are European. They confuse digital with Silicon Valley as if all digital companies come from there.

Europe also has a number of countries taking the lead in the digital transformation, such as the Scandinavian countries, the Benelux, and Ireland. Their digitalisation rate is the highest in the world.

A need for a plan

In the 1980s, Flemish Minister-President Gaston Geens launched the best recovery plan that Flanders (the largest region in Belgium) had ever known: the FTIR plan. One of the 'Third Industrial Revolution in Flanders' spearheads was *Flanders Technology International,* a biannual technology fair that reached the general public. The impact of those fairs, which focused on biotechnology and microelectronics, among other things, can be felt to this day, since Flanders is one of the most successful regions of Europe.

Launching a new plan should be as inspiring as *Flanders Technology International* was back in its day. In addition, everyone, including regions, countries, and European authorities, must be involved. Then comes the difficult step: implementation. This is only possible by making the plan as concrete and public as possible. Make the intended results as measurable as possible so that they are not just empty words. In addition, work in fixed sprints to achieve specific goals. Report what was achieved in the last sprint and what the focus will be in the next on a regular basis.

Overcoming the Covid crisis is a unique time in our history to work on Europe's digital plans, because 'crises are challenges', as entrepreneur André Leysen once said. For a moment, we forget that, under normal circumstances, we're each living on our own island – be it a region, sector, or university. We are facing more or less in the same direction now, as we all want to get out of the crisis as soon as possible, which is why there is a need for a big and inspiring plan.

It's tempting to come up with an all inclusive, comprehensive plan, but that's not how the world of today works. Due to globalisation and digitalisation, it's a lot more complex compared with fifty years ago, which isn't necessarily a bad thing in and of itself. Humans can handle some complexity and that complexity has allowed us to live in a modern and prosperous society. Yet that same complexity also means that we have to hand over some of our control because no one can devise and execute a comprehensive plan by themselves.

It's important that we start now and not wait until 2030. Even if we do nothing, Europe will undoubtedly be fully digitalised by then in order to keep up with other countries, but it still won't be soon enough and perhaps not according to our own standards and values.

If we take the lead in digitalisation, we can set the rules ourselves. Even in the business world, there are still managers who underestimate the advent of digitalisation and how important is it to recognise its magnitude: from Microsoft's Steve Ballmer, who didn't believe in the potential of smartphones, to European entrepreneurs who didn't see in time that technology would also have an impact on their business.

Communicate with the outside world

Midway through her first term in office, New Zealand Prime Minister Jacinda Ardern published a two-minute video detailing all of her government's achievements, including more than 90,000 new jobs and 2,200 social housing projects, the lowest unemployment rate in 11 years, and increased income for 384,000 families. The video was widely shared on social media in no time.

'The obvious response is that New Zealand under Ardern is a super-efficiently governed country, where society is moving forward at breakneck speed. You would think the contrast with Belgium couldn't be greater,' concluded economic journalist Ruben Mooijman of *De Standaard*.

He then went on to see what then Belgian Prime Minister Charles Michel (now president of the European Council) had achieved, and as it turned out, we did at least as well as New Zealand in almost every area! The big difference is that Jacinda Ardern managed to communicate that to the population in a positive way. I think this is typical for Europe: we have prosperous regions with unique values and great culture, but Europeans are no longer enthusiastic about it and instead wonder what the point of Europe even is.

Europe is at the risk of becoming one large open-air museum. We need to reinvent ourselves, as many companies have reinvented themselves at some point. Sometimes they have done this to stay ahead of the competition. A good example is Netflix, which started out sending DVDs through the mail and made the jump to online streaming just in time. Other companies had their backs against the wall when they had to reinvent themselves. A well-known example is Lego, which almost went bankrupt at the turn of the century. By refocusing on their essential message, the joy of building, as well as digitalising, the company was able to once again be a global leader. We must do the same with all of our European companies, but also with Europe itself, as well as each of its member states. When you reinvent yourself, you have to go all the way. That leap of faith can seem scary, but if you don't believe in yourself and your plans, it will have consequences.

Regulate and invest

A government can take on several tasks to digitalise a country, but the most important is that it has the power to create the ground rules that will determine which boat we'll take and which one we'll miss.

In order to keep up to date, the regulations will have to be regularly adjusted. In addition, they will have to be applied correctly. To find a good example of how things can go wrong in this area, we have to go back a hundred years to when the radio was just invented.

The devices were there, but there was no market for them yet. That's why the Belgian companies that developed them started broadcasting radio programmes themselves so that people would buy their devices. The governments of that time let it run its course for ten years until they took drastic action and only allowed the government to broadcast. There was a deep fear that anyone could speak to the population without a filter. Belgium could have become the capital of radio, but after too many restrictive regulations, the innovative technology fell into the hands of a single public company, and the dynamics came to a halt. We must be careful that Europe doesn't do the same with digitalisation. Regulations must not only be up to date and applied correctly; they must also be practicable. For example, the GDPR is fantastic on paper because it protects privacy, but, in practice, it has mainly benefited consultants and lawyers. For many small businesses, its rollout was costly and not a pleasant experience for the user. Or, as professor of information law Patrick Van Eecke put it: 'You can actually summarise the GDPR in one sentence: when you collect personal data from someone else, you must do so with respect. And you can summarise what that respect means in one paragraph. What have we Europeans done? We've written 99 articles to explain those principles, with detailed explanations on how to do it, accompanied by 200 additional paragraphs explaining how to read those ninety-nine articles. And then the national privacy authorities, the DPAS, come together with guidelines on how you should actually interpret it all. It has become a tangled web of regulations, and the only real winners are the lawyers.'

Another important task of the government is to be smart with funding. They need to do more than simply maintain existing systems; they need to prepare us for the future.

Work together
We can also make the 'European internet' better. In America, the World Wide Web looks different than in China and it emphasises different aspects. In the US, technology is mainly

used to trade and create added economic value. In China, it's usually used as a government apparatus. Europe can take the initiative for an internet in which the individual is central and in which cultural norms and values such as privacy, culture, and justice have to be integrated into the core.

The small countries will have to increase their influence within Europe and work together with others to move forward. That is a must for them because their internal markets are too small to be successful on their own when compared with those of Germany and France. Many small European countries, such as Finland, Estonia, and Denmark, have been digital forerunners. They must work together as one unit to work on the digitalisation of Europe, not against superpowers like Germany and France, but as one positive project, because Europe needs it.

The moral of the story is that it's better to take small steps towards great ambitions than to avoid taking an unambitious step because it's too big. And, when that step has been taken, communicate it in a grand way.

Time to get into action

In this book, I have discussed several fears, each and every one of them a human reaction to changing situations. Because we don't know what something is, we tend to resist it. All too often, we assume the pessimistic scenario in which we'll just have to endure the arrival of innovations, but hopefully, with this book, I have shown you that both the problem and the solution lie with us, all with the help of technology.

Technology is neutral. It's about what we do with it. We can use technology to build a control state like Big Brother, but we can also use it to provide everyone with education or to improve our healthcare in an affordable way.

We are at a tipping point in history. Later, we'll look back at this moment and hopefully conclude that the 2020s were the decade in which homo digitalis took matters into their own hands. Maybe this book has been a wake-up call or perhaps confirmation that you're already on the right track. If we want to work on a positive future – and I hope you'll work towards this as well – we can certainly do it.

We are at a tipping point in history: will populisme arise, or will we use technology to build a better world?

Europe needs a new 'Moonshot', an inspiring plan.

1 + 1 = 3

'The network always wins', Peter Hinssen wrote. This book was created, supplemented, and refined by an entire network of friends, acquaintances, colleagues, and relatives. I won't be able to list everybody, but I want to thank them all!

I also thank the 25,000 buyers and readers of my first book, *Digitalis*. It exceeded my wildest dreams and those of the publisher. They were, of course, the best motivation to release this second book. Thank you for the rich and frequent feedback you have sent, which has helped make this book more relevant.

Special thanks to Thomas Smolders, who edited this book. He is particularly good at simplifying and interpreting complex concepts and has been able to process the immense input of all participants with incredible patience and passion.

Ann-Sophie De Steur, for her part, provided the beautiful illustrations in the book, which brought the whole thing to life. Aurélie Matthys of Studio Lannoo then laid out the text and images beautifully.

Thank you to the publishers of Lannoo and Racine, especially Laura Lannoo, Maarten Van Steenbergen, and

Pieter De Messemaeker, who believed in *Digitalis* and encouraged me to write a sequel. Thanks to Stefanie De Craemer for her daily support in operations and marketing.

Many thanks to the readers and experts who gave me input or challenged me. These include:

Alain Gerlache, Andy Serdons, Annie Deweerdt, Babs van Gent, Caroline Coesemans, Caroline Pauwels, Cathy Kremer, Christy Ha, Evelyne Bosteels, François Gilson, Frederik Dooms, Kamal Kharmach, Karl Tuyls, Katya Degrieck, Laurent Hublet, Lola Geerts, Louis Geerts, Nelly Geerts, Manon Geerts, Michiel Sallaets, Philippe Van den Heeden, Pierre Geerts, Stephanie Kaup, Thierry Vermeire, Tom De Block, and Yasmien Naciri.

The entire proceeds of this book's copyrights go to the non-profit association BeCode, about which more is written below.

BeCode non-profit organisation

Since 2016, Karen Boers and company have proven that people can have a significant impact with her BeCode organisation. It offers inclusive training to teach job seekers to program. BeCode has already been able to provide a training course to over 1,500 students, and more than 80% of graduates are able to find work as developers in the IT sector without ever having written a single letter of code before. There are also courses to become a web developer, data AI developer, or security professional, among other things. Motivation and life experience are more important than diplomas in the selection process. I'm very happy that I'm able to contribute the copyright of this book, just like my previous book, in full to BeCode. This way, the organisation can continue to help talented people find a job in the coming years and ensure that companies continue to find the necessary workers to digitalise.

People like Rafah Alani, for example, an Iraqi mother of two daughters. 'I discovered BeCode when I was looking for a job and heard about the course,' she says. 'Every day, there was a new challenge and I learned to work in a realistic working environment. Thanks to BeCode, I was eventually able to work at Lanark as a web developer. Thanks to my coach, Koen, it was the perfect preparation, and I also got to know my best friend there!'

Jeroen De Vetter is also an alumnus of BeCode. He initially trained as a workshop carpenter, but an accident threw a spanner in the works when he was sixteen. 'I almost lost two fingers in an accident with a saw. Then I decided to learn to program and came into contact with BeCode. The internship, in particular, helped me a lot, and because of it, I'm now working at StriveCloud!'

BeCode now also has campuses in Antwerp, Brussels, Ghent, Liège and Charleroi. Do you want to know more? Then go to www.becode.org.